W9-BBF-487

Praise for *The Fight for Climate after Covid-19*

"It is easy to say that the time to act on climate is now. What Alice Hill has done in this thoughtful and informed volume is clearly assess where we can best make decisions to help our governments, our businesses, and our most vulnerable populations adapt to a warmer and higher risk planet."

—SYLVIA M. BURWELL, President of American University
and former Secretary of the US Department
of Health and Human Services

"As Alice Hill explains, there is no shortage of actions that governments, communities, and individuals can take to reduce climate risk. *The Fight for Climate after COVID-19* is a must-read for citizens and policy makers interested in building a future resilient to the next disaster."

—CARLOS CURBELO, former US Representative, Florida

"This visionary page-turner offers a policy roadmap for nations and communities to save lives and improve well-being by adapting to our climate change future, even as we work to control it."

—LINDA FRIED, Dean of the Mailman School of Public
Health and Senior Vice President,
Columbia University Medical Center

"In *The Fight for Climate after COVID-19*, Alice Hill builds the case for climate adaptation in a clear and engaging way, including useful parallels and lessons from the coronavirus pandemic. This insightful book provides a compelling global look at the climate crisis for decision makers at all levels."

—PATRICIA FULLER, Ambassador for Climate Change, Canada

"Alice Hill brings a long lens of scholarly and federal leadership experience to encompass the full landscape of our concurrent disasters, COVID-19 and climate change, and the cascading smaller disasters that have accelerated in frequency and severity over the last several years. *The Fight for Climate after COVID-19* distills her incredible experience and knowledge into insights that shine a light on the way forward to assure the health and welfare of humankind in the context of an ailing global system."

—LYNN R. GOLDMAN, Dean of the Milken Institute School of
Public Health, George Washington University

"Alice Hill's new book reads like an informal briefing in the White House Situation Room on how to handle the climate crisis: Hill knows all the players, all the science, and all the politics. This is an excellent guidebook for any policy maker or citizen of planet Earth who wants to turn this crisis into opportunity and help build a better world."

—JEFF GOODELL, author of *The Water Will Come*

"Coronavirus and climate change have more in common than you might imagine. Experts have warned us of their risks for years, yet as Alice Hill explains, both largely took us by surprise — and that lack of preparation is costing us, dearly. Hill makes an unarguable case with clear, lucid arguments and riveting real-life examples that will stick in your head long after you've put the book down. Anyone concerned about the future must plan for a warming world, and this book shows us how. Essential reading!"

—KATHARINE HAYHOE, Chief Scientist at The Nature Conservancy and United National Champion of the Earth

"While we race to cut emissions to stabilize our climate, we must also prepare for the climate impacts we can't avoid. This book is a must-read for policy makers, businesses, and community leaders looking to secure a prosperous future in a changing climate."

—FRED KRUPP, President, Environmental Defense Fund

"While we must do everything in our power to reduce greenhouse gas accumulation, Alice Hill reminds us that there is no vaccine for the disastrous impacts of climate change. Her book is an invaluable guide to how governments and societies can and must adapt to our rapidly changing climate."

—DAVID MICHAELS, member of the Biden-Harris Transition COVID-19 Advisory Board and former Assistant Secretary of Labor for the Occupational Safety and Health Administration

"Alice Hill provides a fascinating perspective from the frontlines of societal and governmental responses to two huge crises—climate change and a global pandemic. The parallels she offers in this book are highly practical lessons which we should all take to heart."

—ANDREW ROSENBERG, Director of the Center for Science and Democracy, Union of Concerned Scientists

"Even if the world does all it can to reduce emissions, the need for much greater resilience in response to the impacts of climate change grows more urgent every day. In this timely and important book, Alice Hill has written a straight-forward, readable guide to help governments, businesses, and communities take action."

—TODD STERN, former US Special Envoy for Climate Change

"Alice Hill offers a new path in a debate currently dominated by false choices. We must mitigate and adapt to address climate change. Blending personal stories with policy discussions, she offers insights and ideas that are forward-looking, creative, and practical. Anyone interested in the future of our climate, our cities, and our country should read this book."

—FRANCIS X. SUAREZ, Mayor of Miami

"This essential volume illustrates the urgency of the present and the importance of climate education to build a new and better future. Climate crisis requires action on all fronts; it's time we should be able to enjoy the benefits of nature, not just its threats."

—IZABELLA TEIXEIRA, former Environment Minister of Brazil

"*The Fight for Climate after COVID-19* offers a deeply researched glimpse of future disasters and a pragmatic assessment of how to prepare for them."

—SHELDON WHITEHOUSE, US Senator, Rhode Island

ALICE C. HILL

THE FIGHT
FOR CLIMATE
AFTER COVID-19

A Council on Foreign Relations Book

OXFORD
UNIVERSITY PRESS

OXFORD
UNIVERSITY PRESS

Oxford University Press is a department of the University of Oxford. It furthers the University's objective of excellence in research, scholarship, and education by publishing worldwide. Oxford is a registered trade mark of Oxford University Press in the UK and certain other countries.

Published in the United States of America by Oxford University Press 198 Madison Avenue, New York, NY 10016, United States of America.

© Oxford University Press 2021

All rights reserved. No part of this publication may be reproduced, stored in a retrieval system, or transmitted, in any form or by any means, without the prior permission in writing of Oxford University Press, or as expressly permitted by law, by license, or under terms agreed with the appropriate reproduction rights organization. Inquiries concerning reproduction outside the scope of the above should be sent to the Rights Department, Oxford University Press, at the address above.

You must not circulate this work in any other form and you must impose this same condition on any acquirer.

Library of Congress Cataloging-in-Publication Data
Names: Hill, Alice C. (Alice Chamberlayne), author.
Title: The fight for climate after COVID-19 / Alice C. Hill.
Description: New York, NY : Oxford University Press, [2021] |
Includes bibliographical references and index.
Identifiers: LCCN 2021014658 (print) | LCCN 2021014659 (ebook) |
ISBN 9780197549704 (hardcover) | ISBN 9780197549728 (epub)
Subjects: LCSH: Climatic changes—Government policy. | Climate change mitigation. | COVID-19 (Disease)—Government policy. | Medical policy. |
Emergency management—Government policy.
Classification: LCC QC903 .H548 2021 (print) |
LCC QC903 (ebook) | DDC 362.1962/414—dc23
LC record available at https://lccn.loc.gov/2021014658
LC ebook record available at https://lccn.loc.gov/2021014659

DOI: 10.1093/oso/9780197549704.001.0001

3 5 7 9 8 6 4 2

Printed by LSC communications, United States of America

The Council on Foreign Relations (CFR) is an independent, non-partisan membership organization, think tank, and publisher dedicated to being a resource for its members, government officials, business executives, journalists, educators and students, civic and religious leaders, and other interested citizens in order to help them better understand the world and the foreign policy choices facing the United States and other countries. Founded in 1921, CFR carries out its mission by maintaining a diverse membership, with special programs to promote interest and develop expertise in the next generation of foreign policy leaders; convening meetings at its headquarters in New York and in Washington, DC, and other cities where senior government officials, members of Congress, global leaders, and prominent thinkers come together with CFR members to discuss and debate major international issues; supporting a Studies Program that fosters independent research, enabling CFR scholars to produce articles, reports, and books and hold roundtables that analyze foreign policy issues and make concrete policy recommendations; publishing *Foreign Affairs,* the preeminent journal on international affairs and U.S. foreign policy; sponsoring Independent Task Forces that produce reports with both findings and policy prescriptions on the most important foreign policy topics; and providing up-to-date information and analysis about world events and American foreign policy on its website, www.cfr.org.

The Council on Foreign Relations takes no institutional positions on policy issues and has no affiliation with the U.S. government. All views expressed in its publications and on its website are the sole responsibility of the author or authors.

For my parents, Peck and Liz Hill, and my sisters, Emmie and Julie

We choose to do more than cope
With climate change
We choose to end it—
We refuse to lose.

Excerpt from the poem *Earthrise*
by Amanda Gorman, U.S. Youth Poet Laureate

CONTENTS

ACKNOWLEDGMENTS

Shortly after COVID-19 began to take hold in the United States, Janine Zacharia, a journalist turned lecturer at Stanford University, emailed me, reminding me of a conversation we had several years earlier. In 2017, she had asked, "What will it take to stop climate change?" My answer: "A pandemic." I was wrong, at least for this pandemic. The temporary drop in emissions caused by lockdowns won't result in the sustained emissions reductions required. But her email spurred me to think more deeply about the connection between these two catastrophic threats. The idea for a book was born.

Further discussions with Council of Foreign Relations (CFR) President Richard Haass and Senior Vice President, Director of Studies, and Maurice R. Greenberg Chair James Lindsay in April 2020 ensued. By early summer 2020, I had begun writing. I completed the manuscript in early 2021. I could not have met such an aggressive schedule without CFR's extraordinary support. In particular, Richard Haass provided important insights that helped shape the book's central thesis. James Lindsay gave thoughtful feedback throughout the process, assisting me in honing the themes. CFR Vice President and Deputy Director of Studies Shannon O'Neill identified corners to be rounded as I polished the final text. In addition, I am profoundly grateful to CFR Research Associate Madeline Babin, who assisted with research and the thousands of other tasks involved in writing, rewriting, and editing a book manuscript. CFR interns Erica Carvell and Mary Collins also provided research support, as did CFR's wonderful librarians. Patrick Costello and Irina Faskianos helped me deepen my understanding about how federal, state, and local government officials think about climate change. I also thank David Rubenstein for his leadership and generosity as chair of the Board of the Council on Foreign Relations. I am honored to hold CFR's David M. Rubenstein Fellowship for Energy and the Environment.

Sarah Humphreville, editor, at Oxford University Press (OUP), who has now guided my thinking, writing, and vision for two books, deserves especially deep thanks. Her insights on publishing, word choice, and tackling big topics have buoyed this project all along the way. She is a joy to work with and immeasurably talented. OUP Editorial Assistant Emma Hodgdon also provided valued assistance in shaping the final volume. Deep thanks to Kate Munroe Daly and John Carey as well. Each read the entire text with sharpened pencils, sharing valuable lessons about the craft of writing.

I am also grateful for cheerleading, wisdom, and editing advice I received as a fellow in the inaugural cohort of the Op Ed Project and Yale University Fellowship on the Climate Crisis. Chloe Angyal, Martha Southgate, and Ashley Zwick, and my fellow Fellows, have helped me find my voice on climate. Their commitment to finding ways to foment change is inspiring.

The talented team of climate experts at the Environmental Defense Board (EDF), on whose board I am honored to serve, have proved a fount of knowledge. Fred Krupp and Mark Heising have taught me a great deal about how strategic vision and fortitude can change the dialogue. EDF staff and board members Len Baker, Kate

Bonzon, Michael Brownstein, Steve Cochran, Erica Cunningham, Lynn Goldman, Charles Hamilton, Nat Keohane, Richard Lazarus, Amanda Leland, Frank Loy, Ray Mabus, Tom Murray, Rinnie Nardone, Susan Oberndorf, Michael Panafil, Vicki Patton, Laura Rodriguez, Mark Rupp, Eric Schwaab, and Natalie Peyronnin Snider, as well as other EDF board members and staffers, have deepened my understanding of the existential threat of climate change.

The Center for Climate and Security and the Council on Strategic Risks has sharpened my appreciation for the growing security threats posed by climate change and biological threats. I am honored to be a part of this dedicated community and am particularly grateful to stalwarts Miranda Ballentine, Bob Barnes, John Conger, Francesco Femia, Sherri Goodman, Deborah Gordon, Kate Guy, Marcus King, Dennis McGinn, Christine Parthemore, Ann Phillips, Erin Sikorsky, David Titley, Andy Weber, and Caitlin Werrell.

This book benefited from the guidance and insights generously shared by many people. To each of them I am greatly indebted. Among them are Newsha Ajami, Thad Allen, Sarah Anderson, Terry Anderson, Paul Angelo, Jeffrey Arnold, Vicki Arroyo, Adrienne Arsht, Paul Auerbach, Alyssa Ayres, Bilal Ayyub, Jack Baker, Noel Bakhtian, John Balbus, Marjorie Baldwin, Greg Barats, Anne Barnard, Josh Barnes, Kit Batten, Jainey Bavishi, Phillippe Benoit, Bidisha Bhatacharyya, Rosina Bierbaum, Steven Bingler, Laura Blatchford, Antony Blinken, Analise Blum, Louis Blumberg, Paul Bodnar, Thomas Bollyky, Robert Bonnie, Thomas Bostick, Francis Bouchard, Cynthia Brady, Kate Brandt, Chad Briggs, Esther Brimmer, Steven Brock, Stan Bronson, Sharon Burke, Mackenzie Burnett, Sylvia Mathews Burwell, Josh Busby, James Butler, Andrea

ACKNOWLEDGMENTS

Cameron, John Campbell, Christina Chang, Stephen Cheney, Anne Choate, Rachel Cleetus, Joel Clement, Michael Coen, Ryan Colker, Heather Conley, Kristina Costa, Carlos Curbelo, Henri de Castries, Roger-Mark De Souza, Todd De Voe, Geoffrey Debalko, Brian Deese, Scott Deitchman, Chloe Demrovsky, Tamara Dickinson, Taylor Dimsdale, Paula Dobransky, Shaun Donovan, Rowan Douglas, Maureen Downie, Daniel Drollete, Phil Duffy, Kerry Duggan, Cheryl Durrant, Jon Eberlan, Jim Edmondson, David Espinoza, Edward Etzkorn, Aaron Ezroj, David Festa, Natalie Anne Festa, Steve Fetter, Shiloh Fetzek, Chris Field, Leslie Field, Christopher Flavelle, Ann Florini, Karen Florini, Stephen Flynn, George Frampton, Robert Francis Jr., Dr. Linda Fried, Craig Fugate, Patricia Fuller, Sarah Gaines, Richard Galant, Gerry Galloway, Michael Gerrard, Daniel Gerstein, Tom Gilligan, Rene Gobonya, Christy Goldfuss, Jeff Goodell, Jessica Grannis, Henry Green, Michael Gremillion, Sean Griffin, Michael Grimm, Lee Gunn, Elizabeth Hadley, Avril Haines, Stephane Hallegatte, Howard Harary, Jan Hartke, David Hayes, Katherine Hayhoe, Lukas Haynes, Carl Hedde, Claude Henry, Victoria Herrmann, Sam Higuchi, Jennifer Hillman, John Holdren, Andrew Holland, Ken Hudnut, Dr. Richard Hunt, Saleemul Huq, Dr. Thomas Inglesby, Katharine Jacobs, Michael Jolicoeur, Chris Jones, Sarah Jordaan, John Jordan, Jennnifer Jurado, Dr. Ali Kahn, Mathew Kahn, William Kakenmaster, Brian Kamoie, Mike Kangior, Daniel Kaniewski, Jesse Keenan, John Kerry, Daniel Kessler, Kelley Kevin, Heather King, Wendell Chris King, Amanda King, Ron Kingham, Avery Kintner, Thomas, Kirsch, Ron Klain, Andreas Kleiner, Jay Koh, Ann Kosmal, Carolyn Kousky, Daniel Kreeger, Anthony Kuczynski, Howard Kunreuther, Michael Kuperberg, Daniel Kurtz-Phelan, Jennifer Kurz, Rachel Kyte, Sarah Labowitz, Loren

ACKNOWLEDGMENTS

Labovitch, Sarah Ladislaw, Olivia Lazard, Amanda Leck, Anthony Leiserowitz, Robert Lempert, Eric Letvin, Laurie Levenson, James Lewis, Andrew Light, Laura Lightbody, Justin Lindsay, Igor Linkov, Glynis Lough, Amy Luers, Dr. Nicole Lurie, Jane Holl Lute, Nick Mabey, Katherine Mach, Brenda Mallory, Felicia Marcus, Jeremy Marcus, Lenny Marcus, Jennifer Pinto Martin, Leonardo Martinez-Diaz, Douglas Mason, Andrew Mayock, K. C. McCarten, Megan McCaslin, Denis McDonough, Mike McGavick, Meg McLaughlin, Kathy Baughman McLeod, Eric McNulty, Monica Medina, Samantha Medlock, David Michaels, Erwan Michel-Kerjan, Lisa Monaco, Rafaela Monchek, Rob Moore, Neil Morisetti, Christine Morris, Stephanie Morrison, John Morton, Richard Moss, Major General Muniruzzaman, Senator Lisa Murkowski, Paul Musgrove, Melanie Nakagawa, Janet Napolitano, Sissy Nikolau, Joe Nimmich, Neil Noronha, Frank Nutter, Kevin O'Pray, Tara O'Toole, Sally Ourieff, Robert Papp, Meaghan Parker, Doug Parsons, Michelle Patron, Eric Pelofsky, Jonathan Pershing, Jeff Peterson, Mike Peterson, Laura Petes, John Podesta, Dr. J. D. Polk, Daniel Poneman, Amy Pope, Tim Profeta, Vinukollu Raghuveer, Andrew Revkin, Susan Rice, T. C. Richmond, Lauren Herzer Risi, William Robbie, Valerie Rockefeller, Amy Rosenband, Cheryl Rosenbaum, Andrew Rosenberg, Lea Rosenblum, Marcie Roth, Arghya Singha Roy, Susan Ruffo, Joel Scata, David Scheffer, Joel Scheraga, Rod Schoonover, George Schultz, Peter Schultz, Mike Sfraga, Julie Shafer, Afreen Siddiqi, A. R. Siders, Erin Sikorsky, Dominic Sims, Nina Sovich, Douglas Smith, Steve Smith, Jon Spaner, Daniel Stander, Todd Stern, Craig Stewart, Skip Stiles, Stephanie Strazisar, Francis Suarez, Stacy Swann, Bianca Taylor, Izabella Teixeira, Lynn Tennefoss, Ronald Thieman, C. Forbes Tomkins, Shana Udvardy, Francis Ulmer, Dan Utech, Lisa Van Dusen, Bina Venkataraman,

Alexander Verbeek, Ahmad Wani, Anne Waple, Michael Wara, Emily Wasley, Nancy Watkins, Caitlin Werrell, Kathleen White, Senator Sheldon Whitehouse, Leslie Williams, Henry Willis, Steven Wilson, Marcia Wong, Sierra Woodruff, Roy Wright, Don Wuebbles, Sally Yozell, Janine Zacharia, Judy Zakreski, Craig Zamuda, Daniel Zarrilli, and Paul Zukunft.

I am also in debt to the anonymous reviewers who saw early drafts of the book proposal and manuscript. The perspectives and examples they provided helped refine my analysis of the parallels between pandemics and climate change, as well as the challenges ahead.

Finally, I want to thank my family, the Hills and the Starrs, for their love and support. My sisters, Emmie and Julie, to whom I have dedicated this book, have been my go-to people for fun and advice all my life. My deepest gratitude goes to those to whom I dedicated my first book: Peter Starr, my college sweetheart who has been supporting and cheering me on since the first day we met, and our two wonderful children, Liza and Julia Starr, who have brought immeasurable love, laughter, and joy into my life. Getting to spend so much time with each of them was this pandemic's silver lining.

INTRODUCTION

Necessity is the mother of adaptation.

—*Angela Duckworth, American author and professor of psychology*

2020 was a year like no other. It brought the planet the worst disaster in living memory, a pandemic that invaded virtually every corner of the earth. The novel coronavirus stole livelihoods and lives around the globe. Governments ordered people to stay home and schools, restaurants, and theaters to shutter. Supply chains convulsed as factories powered down assembly lines, global shipping plunged, and planes sat idle on tarmacs. Hotels went begging for guests. Whole industries went into a tailspin. Companies reversed course, telling employees not to come to the office but to work from their kitchen tables instead. Commercial real estate suffered massive vacancies. With jobs lost in droves, worldwide hunger spiked, as did domestic violence, homicides, and child marriage. Meanwhile governments pumped out trillions of dollars in stimulus packages to support their struggling economies and populations. A whole new vocabulary relating to disease erupted. Words like "lockdown," "social distancing," "reopening," "superspreader," and "COVID-19" shot into the English-speaking world at hyper-speed, according to the *Oxford English Dictionary*.

But it wasn't just the worldwide pandemic that caused 2020 to be like no other year. In 2020, climate change, once considered a threat for the distant future, arrived with a vengeance, cutting its own swath of destruction across the planet. The year tied with 2016 for the hottest on record. Temperatures soared to likely the highest ever recorded in the Arctic Circle, 38 degrees Celsius (100.4 degrees Fahrenheit), as well as the highest ever reliably recorded on earth, 54.4 degrees Celsius (130 degrees Fahrenheit) in the aptly named Death Valley in California. Historic wildfires consumed so much land in the western United States, over ten million acres (4.04 million hectares), that a meteorologist at the U.S. National Interagency Fire Center labeled the damage "surreal." Explosive bush fires in Australia killed three billion animals. "Zombie fires" smoldered through the long Arctic winter in Siberia.

The strongest cyclone ever to make landfall hit the Philippines, carrying record-breaking sustained wind speeds of 195 miles per hour (314 kilometers per hour). Pakistan experienced its wettest August on record, and on a single day in August, 9 inches (231 millimeters) of rain fell on Karachi, the highest daily total ever recorded. The melt of global land ice continued at breakneck speed, driving sea levels to rise ever higher. The Atlantic hurricane season carried so many named storms that meteorologists turned to the Greek alphabet to come up with new names. In the Horn of Africa, 200 billion locusts flew in voracious swarms 20 times the size of Paris, devouring 50 to 80 percent of crops in the field. That's an estimated 8,000 times more locusts than would appear in the absence of climate change and the equivalent of 25 for every person on the planet.[1]

2020's string of record-breaking events also carried a record-breaking invoice for damages. According to the global reinsurance company Munich Re, the economic losses from natural disasters jumped from $166 billion in 2019 to $210 billion in 2020.[2] In response to the mounting costs, one member of Munich Re's board urged, "It is time to act."[3]

My climate education

I never set out to work on climate change. Instead, like many in the field, I simply got a climate assignment one day and never let go. My route to the issue was admittedly circuitous. I got my start in law, first as a federal prosecutor in California. When Charles Keating Jr. and his son committed the biggest bank fraud ever uncovered by the Federal Bureau of Investigation up to that time, I was co-lead prosecutor on the case. I eventually headed the white-collar crime prosecution unit at the U.S. Attorney's Office in Los Angeles. California Governor Pete Wilson, a Republican, later appointed me to the California state judiciary. As a judge, I adjudicated thousands of cases ranging from gang-related homicides to medical malpractice and served as supervising judge on both the municipal and superior courts.

Shortly after the election of Barack Obama as president in November 2008, my phone rang. On the call was Janet Napolitano, a friend from law school, asking, "How would you like to come to Washington?" That moment yielded one of the most important pieces of career advice I can share: "Be nice to those you sit next to in school." Obama had recruited Napolitano to become secretary of the Department of Homeland Security (DHS), the third largest agency in the U.S. government after the Departments of Defense and Veterans Affairs. DHS had been born out of the terrorist attacks of September 11, 2001, in the biggest reorganization of the federal government since World War II. Napolitano wanted me to help her manage the young, sprawling agency.

As senior counselor to Secretary Napolitano for four years, I served as jack-of-all-trades at DHS, overseeing preparedness and response to biological threats, including pandemics; creating the department's first-ever climate adaptation plans; and founding an anti–human trafficking initiative, the Blue Campaign, among other things. I also served on the committee responsible for producing the country's national climate assessment, a report mandated by the U.S. Congress. When Napolitano left Washington to become president of the University of California system, I joined the White House. I eventually became special assistant to President Obama and senior director for resilience policy on the National Security Council (NSC), the president's principal forum for making decisions about national security, foreign policy, and military matters. At the NSC, I led the Resilience Policy Directorate, supervising a team of NSC directors, reporting to Homeland Security Advisor Lisa Monaco as well as National Security Advisor Susan Rice and her deputies, Tony Blinken and Avril Haines.

The Resilience Policy Directorate had offices in an ornate suite with 12-foot ceilings, carved moldings, and shaded windows on the fourth floor of the Old Executive Office Building, which sits only a driveway away from the Oval Office. Our job was to develop national policy aimed at reducing catastrophic risk of national consequence. The talented team of directors included emergency managers, public health experts, infrastructure experts, lawyers, engineers, military officers, and even an emergency room doctor. Our responsibilities encompassed policy oversight for natural disasters, biological and chemical threats, public-private partnerships, critical infrastructure, and climate change.

During my tenure as senior director, the NSC Resilience Policy Directorate broke new policy ground on climate change for the nation. We created the country's first national building standards for the climate-worsened risks of flood and wildfire, as well as earthquakes. We worked with then–Vice President Joe Biden to pioneer a dialogue among the country's fire department chiefs to identify better ways to combat increasingly destructive wildfires. When the Ebola epidemic struck, some of my team joined Ron Klain's Ebola Task Force to stop the spread of the disease in the United States. Others worked to curtail antimicrobial resistance as well as build biological threat preparedness. We launched a partnership with the deans of public health, medical, and nursing schools to spur greater education regarding climate-worsened health risks. We developed strategies for addressing climate-driven changes in the Arctic, an area that is warming at least twice as fast as the rest of the planet, causing a new ocean to open for navigation. We devised a fresh approach for providing much-needed government support to drought-stricken communities. We crafted executive orders signed by President Obama focused on managing national security risks from climate change, ranging from preparing military bases for sea-level rise to making overseas investments "climate smart" by determining whether the proposed new road in a developing country could withstand future climate-worsened flooding.

When I finally left the White House in 2016, Susan Rice joked that if I hadn't already decided to leave on my own, she would have had to shut my directorate down—the rest of the White House couldn't keep up with our volume of work. My team, according to Lisa Monaco, had broken the record for executive orders signed by President Obama.

While at the White House, I also served as a member of Obama's climate team, first led by Counselor to the President John Podesta and later by Brian Deese. That role took me to Anchorage, Alaska, to serve as a U.S. delegate for the international GLACIER (Global Leadership in the Arctic: Cooperation, Innovation, Engagement, and Resilience) conference organized by then-U.S. secretary of state John Kerry in the summer of 2015. In December 2015, I traveled to Paris for the UN Conference of Parties that resulted in the historic Paris climate accord.

Working in the Obama administration on the catastrophic risks posed by climate change, and then later at the Hoover Institution at Stanford University and at the Council on Foreign Relations, has given me the opportunity to learn from Nobel Prize–winning climate scientists, MacArthur genius grant awardees, meteorologists, state and local resilience leaders, owners and operators of critical infrastructure, architects and engineers, developers, youth activists, environmentalists, educators, journalists, lawyers, statisticians, actuaries, catastrophe modelers, military leaders, diplomats, intelligence analysts, political scientists, management consultants, insurers, emergency managers, humanitarian aid providers, economists, city planners, bankers, financial system regulators, public health officials, doctors, and dozens of elected politicians. These conversations have divulged the nuance and uncertainty that can accompany decisions about climate change. They have also revealed that far too many decision makers still discount the risks from climate that lie ahead.

Discounting climate risk

Unfortunately, many people within government and business, like me, have never received any formal education about climate change, either in their professional lives or in school. Nor have they had the opportunity that I have had to immerse myself in the topic. What they know, they have typically gleaned from news reports and conversations. This lack of training, or what some call "climate literacy," adds to the deep political divide that cleaves discussions about what to do about climate risk, with some leaders even questioning whether climate change is occurring. The resulting patchwork in understanding has meant that many decision makers simply do not yet appreciate either the exponential nature of the changes wrought by temperature rise or how those changes will undermine every human-created and natural system.

Because the mounting risk remains veiled to some, they tend to discount it. This can make developing policy for climate change at times feel like pushing a boulder uphill. Even some of those working for President Obama, whose administration was the first to make climate change a priority, did not appreciate what was at stake. This was brought home to me shortly after I arrived at the White House to join the NSC staff.

The federal government had just issued its *Third National Climate Assessment*, a report that made clear the country would face more severe flooding, bigger wildfires, and greater heat extremes in the near future. I had called a meeting with federal emergency management officials to discuss how the government could assist American communities in preparing for climate-worsened disasters predicted by the report. The officials, however, told me and my team that doing more to prepare for climate change was not a priority. Several months later, I was sitting in my office in the Old Executive Office Building when a senior director on the NSC stopped by. As he stood in the doorway, he warned that I was "getting a reputation." "For what?" I asked. "For worrying too much about climate change," he replied. These and other encounters made plain to me that some policymakers still do not appreciate how the accumulation of greenhouse gases will affect the planet as we know it.

In addition to a lack of education, the fact that the human brain tends to assess risk based on recent experience adds to the formidable barrier to understanding the escalation of climate threats. We prioritize risks that we can easily recall and discount those with which we are unfamiliar or consider unlikely to happen. Our life experience teaches us that the future generally will resemble the past; the next flood will resemble the last flood, and so on. As a consequence, we fail to imagine the intensification of extreme events as global temperatures rise. With climate change, however, clinging to the belief that the future will resemble the past leads to poor choices that leave people increasingly vulnerable. With climate change, what matters most is what might come next: the past is no longer a safe guide for the future. Leaders, as well as ordinary citizens, need to pay attention now to what the climate science predicts the future will hold. They need to plan for potential breakdowns of systems on a scale even beyond what the pandemic has caused.

This lesson, as it turns out, is a hard one for us humans to grasp.

Mitigation is no longer enough

2020, in addition to bringing the pandemic and relentless weather disasters, brought another grim new record for the planet—the amount of carbon dioxide (CO_2) in the atmosphere. For the past 6,000 years or so, the earth has enjoyed a stable climate in which human civilization has bloomed. But, starting around 1850, the Industrial Revolution sowed the seeds of future change. Even as the burning of coal, oil, and natural gas brought unprecedented prosperity and progress, it also spewed increasing amounts of heat-trapping gases such as CO_2 into the atmosphere.

In 1958, scientists at Mauna Loa Observatory, which sits 11,145 feet (3,397 meters) above sea level in Hawaii, began to track the level of carbon emissions in the air. By then, the amount of CO_2 had reached 315 parts per million (ppm), up from an estimated concentration of 280 ppm during the preindustrial era. Ever since, CO_2 levels have continued their upward climb. The rate at which carbon emissions accumulate is accelerating. In the 1960s, the annual growth rate averaged about 0.8 ppm. By the 1980s, that rate had doubled, and in the 2010s, it climbed to 2.4 ppm per year.

The accumulation of gases in the atmosphere traps heat on the earth like the glass roof of a greenhouse. The planet is now warmer than it has been during the entire history of human civilization. With more greenhouse gases piling up, the heating will increase, which, in turn, will fuel bigger weather extremes that will pummel every corner of the planet (Figure I.1).

In 1992, the United Nations convened the world's countries in Rio de Janeiro, Brazil, to act on climate. Countries signed on to the UN Framework Convention on Climate Change (UNFCCC), which aimed to prevent "dangerous" human interference in the climate system. It called on nations to reduce greenhouse gas emissions, which is also known as "mitigation." The UNFCCC did not include any legally binding mitigation requirements. But it did require frequent meetings among nations, namely the Conference

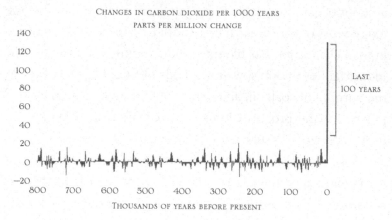

FIGURE I.I. Current and past atmospheric carbon dioxide levels, with recent increases caused primarily by the burning of fossil fuels.
Source: Climate Central.

of the Parties, or COP, which the UN has hosted ever since. In 1997, the COP held in Japan resulted in the Kyoto Protocol, which required developed countries to cut emissions. But the world's then-largest carbon emitter, the United States, pulled out less than four years later, claiming that the protocol was not in its economic best interests. Global action on climate would have to wait.

In December 2015, international representatives met again, this time in Paris for COP 21. By then, countries had abandoned the goal of mandatory reductions and instead agreed to make voluntary "nationally determined contributions" (NDCs) to reduce greenhouse gas emissions. Their aim was to keep temperature rise below 1.5 degrees Celsius (2.7 degrees Fahrenheit) above preindustrial levels. They further promised to provide revised NDCs by 2020. But before nations could reconvene, the pandemic struck, prompting UN leaders to postpone the reckoning on NDCs until 2021. Meanwhile, emissions continued to mount.

By mid-2020, the level of CO_2 emissions ratcheted up to 418 ppm, the highest level in human history and likely the highest since three million years ago when sea levels were 60 feet (18 meters) above those today.

Meanwhile, in advance of the approaching 2021 COP, many countries admirably promised to increase their ambition to cut carbon, becoming carbon neutral or net zero within three or four decades. But because past emissions of heat-trapping gases now wrap around the earth like a thick blanket, temperatures will continue to rise for the foreseeable future. So, even if countries honor their pledges to mitigate, which for the sake of the health of the planet they absolutely should, the earth will continue to warm. This means that governments, businesses, communities, nongovernmental organizations (NGOs), philanthropies, and individuals should act now to prepare for and respond to new climate-fueled extremes, like deeper droughts, extreme precipitation that falls in "rain bombs," bigger wildfires, more intense storms, new heat extremes, and relentless sea-level rise. And if mitigation efforts fall short of the 1.5-degree-Celsius goal, the danger will grow greater.

In other words, mitigation alone will no longer keep us safe. We also need to adapt.

Adaptation needs a seat at the table

For decades, some experts in the NGO, scientific, academic, and policymaking community treated any discussion of climate adaptation, or what has more recently been termed "resilience," as taboo. They reasoned that since climate impacts would occur in the distant future, adaptation could wait until that future arrived.

Mitigation, on the other hand, demanded immediate attention to attack climate change at its source. They feared that talking about adaptation would send a false signal, seducing government and private sector leaders into assuming that adaptation alone could solve the climate problem and that cutting emissions was not urgent or even necessary.

Or worse yet, embracing adaptation might be viewed as "an admission of failure on mitigation."[4] Researchers who chose to study adaptation were derided as "defeatist" and "fatalist," with adaptation treated as a "poor cousin" to mitigation.[5] Not surprisingly, the two sides of the climate coin didn't "talk to each other," said one French scientist.[6] The divide even stretched into the legal community, with American legal scholarship focused mostly on mitigation strategies, such as comparing the merits of cap-and-trade to carbon taxes.

That division has inhibited each group from sharing insights and objectives culled from the other camp among themselves or with the public. The divide has also meant that mitigation of emissions has dominated the climate policy discussion, with adaptation continuing to play "second fiddle" to efforts to reduce global greenhouse gases.[7] In January 2021, the continued focus on mitigation caused UN Secretary-General António Guterres to call for "a breakthrough on adaptation." He warned, "Adaptation cannot be the neglected half of the climate equation." Yet, just a few months later, when newly-elected U.S. President Joe Biden convened an international climate summit in Washington on the 51st anniversary of Earth Day, his administration's proposed climate actions tilted toward mitigation. With adaptation often lacking a seat at the table, it has remained undervalued by governments, think tanks, corporate boards, and financial and academic institutions.

Fortunately, the gulf between mitigation and adaptation efforts has begun to close in some areas recently, but its effects persist. One way to view this is through funding. A 2019 analysis estimated that mitigation activities captured 93 percent of global financing from public, private, and other sources in 2017–2018.[8] Adaptation's share amounted to just 5 percent, with almost all of that from the public sector, while the remaining 2 percent went to combined adaptation and mitigation activities. Similarly, researchers in Sweden studied 14 international organizations, including the World Bank, major regional organizations, and the United Nations, and found that those institutions had reduced the average amount of funding provided for adaptation in the decade from 2008 to 2017, as compared to 1999 to 2007.[9]

We can no longer hope to preserve the climate we have known. We will have to adapt.

Adaptation requires action on all fronts

Climate change has already begun to disrupt virtually every system, be it wreaking havoc on naturally occurring events, such as seasonal rainfall, or human-made infrastructure, like the electric grid during wildfires. To adapt to anticipated climate risks, governments, communities, businesses, and individuals need to embark on making widespread, foundational adjustments to how and where they live and conduct business. They can no longer assume that decisions they made in the past about construction standards, land use, disease surveillance, transportation systems, power generation, water access, flood protection, emergency management, wastewater treatment, supply chain integrity, or other critical matters will keep them safe going forward. Planning and preparing for climate-worsened extremes requires action along all axes of government and sectors of society.

To make sure that those decisions work in concert, national governments should set the framework. They need to provide the strategic vision for achieving long-lasting resilience. From that strategy, adaptation at the state, local, and private levels should flow. But adaptation planning around the world remains highly fractured, involving, for example, decisions within interior ministries about support for development in risky areas, within foreign ministries about repurposing development assistance to help the least developed nations cope with their growing challenges, and within finance ministries regarding budgetary allocations for burgeoning disaster costs. Without coordination, adaptation becomes piecemeal, watered down, and no match for the scale of action required.

Many adaptation decision points will first land at the local level since that is where climate impacts first fall. These decisions remain geographically specific as they relate to the particular vulnerability of an area to future climate impacts. They also remain highly politicized. Where the local road will be rerouted, or which houses get moved to make way for flood protection, matters deeply to those affected. In the absence of a clear national vision that seeks coordination among internal and international neighbors, local adaptation efforts risk becoming a series of Band-Aid fixes with no lasting power.

Even when adaptation does take place, governments and businesses remain ill-prepared to track progress and share lessons learned. Policymaking for adaptation suffers from the lack of a common, internationally accepted metric as simple as that for mitigation, the number of metric tons of greenhouse gases emitted into the atmosphere. Therefore, the collection of data regarding adaptation, as well as the tracking of the implementation and efficacy of particular adaptation measures, has proven messy. As nations and communities embrace adaptation, assessing and aggregating progress across multiple dimensions, countries, levels of government, and sectors can feel as daunting as curing the common cold.

Spurring climate adaptation

Adaptation, when it does occur, often happens after what my coauthor, Leonardo Martinez-Diaz, and I termed "no more" moments in our book, *Building a Resilient Tomorrow*. "No more" moments take advantage of humans' tendency to assess the probability of an event based on how easily an example of the event comes to mind. That's why, for example, immediately after a flood, homeowners become more interested in flood-proofing and communities mobilize to reduce future flood risk. Similarly, there is no doubt that the worldwide devastation caused by the coronavirus pandemic has provided a "no more" moment that will lead to improved global public health preparedness.

But with climate change there is no single "no more" moment to galvanize public support and thus political will. The very nature of climate change dulls the sense of urgency. Because climate impacts manifest in different locations at different times in seemingly unrelated ways, they can seem episodic to the casual observer. Their variety, temporal variation, and geographic sprawl act to disguise their global cumulative growth. By the time the public finally coalesces around action, communities could face unmanageable risk, risks that could prove wildly expensive to address or even dangerous, as in the case of geoengineering that intervenes to alter the climatic system. As warming continues, the current global reluctance to having planes make daily flights depositing sulfate particles in the sky to reflect solar radiation back to space, for example, could seem like a safer bet than continued runaway heating. But no matter whether nations or billionaires take action to rejigger the climate, everyone will have to adapt to climate change given the environmental changes that are already "baked in" due to past emissions.

No vaccine exists to solve climate change. There is no silver bullet. Given the urgency for action and the costs entailed, it's important that countries and communities alike make smart choices about mitigation and adaptation that work together. Decision makers must evaluate emissions reduction and preparedness policies with an eye to their overall contributions to addressing climate risk. Neglecting to consider these solutions as they relate to one another runs the added risk of worsening the climate crisis.

The pandemic has sadly provided a "dress rehearsal" for nations, communities, and individuals in confronting catastrophic risk. That rehearsal has revealed valuable lessons, including on how to adapt more effectively to climate change. Those lessons can drive better outcomes as temperatures continue to climb. First among them is that preparation and early action matter. For both climate change and pandemics, an ounce of prevention will yield at least a pound of cure.

Yet, preparedness lags. It lags at all levels of government, from national to local, as well as in the private sector. This is true even though the world has never had better foresight. As my years working alongside public health and climate experts taught me, putting that foresight to use would reduce suffering, while providing a competitive advantage. With no single crisis moment driving action, the risk grows that the collective action needed to tackle the problem of reducing emissions will not happen in time to avoid catastrophic heating. That means that adaptation deserves close attention now more than ever. And that's why I wrote this book.

Getting started on adaptation

If we make better choices today, we can help buffer future climate blows. Pathways to improving our decisions around climate risk already exist, as do guideposts for securing greater preparedness and resilience in the future. By illuminating the way forward, my aim is to give decision makers ideas for where to get started on some of the hardest work of all: determining how to thrive in a warming world.

In the first chapter, I map the essential terrain for understanding climate risk: the future will no longer resemble the past. This singular characteristic of dealing with climate change poses a powerful barrier to successful adaptation. Unlike infectious disease, global warming causes changes to the environment that can occur at a magnitude and speed never experienced within human history. To get started on accounting for a future with worsening extremes, nations, businesses, and communities need to get serious about planning.

That planning should consider the quickening pace of crises that climate change will bring and how extremes striking in rapid succession will compound the ensuing damage. All of this makes it even more difficult for people and communities to climb out from the rubble left behind. Weakened by simultaneous catastrophes, ranging from cyclones to droughts, governments will struggle to contain criminal opportunists, including terrorists and organized crime, as the world has witnessed during the spread of the coronavirus. In Chapter 2, I explore how reinforcing and amplifying emergency preparedness could blunt the trauma that climate impacts spread.

The pandemic serves as a warning that catastrophic risks do not honor geopolitical or jurisdictional boundaries. Building resilience to these risks involves planning regionally and across long-standing borders. Chapter 3 looks at promising regional cooperation efforts to de-escalate tensions heightened by climate change, ranging from expanding insurance availability to stockpiling and re-energizing multilateralism.

Climate change impacts, like pandemics, increase human vulnerability to harm. Preparedness efforts should, therefore, attend closely to the needs of those who are at greatest risk when disaster strikes. In Chapter 4, I take stock of promising ways to protect vulnerable populations in the face of worsening extremes, including providing relief from unsustainable sovereign debt and reducing gender inequality.

For those communities that can manage to muster the political will to act in the absence of "no more" moments, I pinpoint the easy wins for building resilience in Chapter 5. Those range from ensuring that all new infrastructure can withstand climate impacts expected over the course of its service life to requiring corporate disclosure of climate risk and commemorating past disasters to keep the "no more" moments alive for future generations.

Finally, in Chapter 6, I scrutinize the consequences of the gnawing disconnect between policies focused on mitigation and those focused on adaptation. The tendency of climate experts to work in silos increases the likelihood of "malmitigation," investments in green energy that are vulnerable to climate impacts, as well as "maladaptation," investments in resilience that contribute to greenhouse gas emissions. If designers and planners neglect to consider both sides of the climate coin, they risk exacerbating harm from climate change.

The truth we must all face, that I've mentioned here already but want to press home again, is that even if every country in the world doubles down on cutting emissions, we have already baked in decades of rising temperatures. The fight for climate change now requires both mitigation *and* adaptation. Within these pages, I have laid out a blueprint to get started on adaptation. Informed by lessons from the pandemic, as well as my DHS and White House experience, I hope to help ordinary citizens, government planners, and business leaders alike to answer the questions: what does climate change mean for me, and what should my community and I do about it? Whether we like it or not, the world will be different than it is now, and we will have to adapt. Read on to find out how that adaptation journey has its best chance of success.

CHAPTER 1

ACCOUNT FOR THE FUTURE, NOT JUST THE PAST

Imagination is more important than knowledge.

—*Albert Einstein*

As the coronavirus began to ricochet across the United States in early 2020, U.S. President Donald Trump wondered out loud, "Who would have thought?"[1] Even as the number of infections soared over the next few weeks, President Trump continued to insist that the pandemic "was unexpected." The United States notoriously stumbled in its response to the spread of the disease, suffering the greatest loss of life of any developed country. For years to come, hearings, investigations, and reports will plumb the depths of government for the causes of this national calamity. Central among the shortcomings uncovered will be "a lack of imagination," according to Dr. Ali Khan, former head of the U.S. Centers for Disease Control and Prevention's Office of Public Health Preparedness and Response.[2]

Devastating pandemics have struck repeatedly throughout human history—from the Black Death of the 14th century to the cholera pandemic of the early 1900s. While advances in public health have completely or largely tamed some of the past's worst scourges, like smallpox and plague, the risks of new pandemics have grown in recent decades. From 1920 to 2020, the percentage of people living in urban areas increased from 20 percent to over 50 percent. With more people crowding on top of each other in cities, disease can spread more easily. Likewise, more people traveling by plane and to farther distances makes it easier for pathogens to cross borders and spread rapidly around the planet.

Human encroachment into forests and other natural spaces means more people are coming into contact with wild animals, creating more chances for disease to jump from animal species to humans. Industrial farming adds to the risk of animal-to-human disease transmission, as do more people living near livestock, visiting wet markets, and consuming wild animals. Since the 1970s, over three dozen infectious diseases have emerged from animal contact with humans, including chikungunya, Ebola, Middle East respiratory syndrome (MERS), severe acute respiratory syndrome (SARS), Zika, and several variants of avian and swine influenza. Seventy percent of all emerging infections are now traced to human-wildlife interactions. As a result of these developments, epidemiologists and public health experts had predicted for years that a pandemic caused by something like the coronavirus was virtually inevitable. Yet when the disease began to race around the globe, communities and nations still acted surprised. They suffered from a failure of imagination.

That failure raises a profound question about humans' ability to cope with future challenges: If imaginations fail even with known risks, how can governments and communities possibly prepare themselves for threats and changes that have never occurred before in human history, and that may not make themselves known for years or decades? What will it take to spur the necessary action in the absence of a "no more" moment?

Foremost among those new threats is climate change. The planet has enjoyed a relatively steady climate for close to 7,000 years, with global temperatures and sea-level rise remaining stable until recently. The assumption that the earth's climate will continue to remain stationary has traditionally guided every decision about how and where humans live. But that assumption is no longer valid. Human activity has caused the earth to warm 20 times faster than when it transitioned out of the last ice age about 12,000 years ago, and that rate is accelerating. These days, every month brings unimagined and unfamiliar extremes, from searing heat waves to deluges on a scale never before experienced in recorded history. To withstand future climate change impacts, communities and nations cannot continue to rely on the past as a safe guide for the future, as they have for thousands of years. Instead, they should imagine the unknown.

To avoid the collective failure of imagination that currently impedes efforts to address climate risks, decision makers, be they political or business leaders, should confront a never-before-experienced reality, namely that the globe's climate has begun to move beyond its stable boundaries. A growing set of climate tools and services allow humans to identify and understand the looming future risks. Employing these tools to plan for climate change on a regional, national, and local scale, as well as within the private sector, will allow communities to build awareness and identify fresh approaches for adapting to the new climate reality. All of the planning in the world, however, will not ensure success when leaders lack the understanding, the will, or the skill to turn scientific information and plans into concrete action. As the fumbled response to the COVID-19 crisis in the United States revealed, leadership matters when it comes to global crises, like pandemics and climate change.

Learning to care about climate change

In October 2009, U.S. President Barack Obama put his foot on the accelerator to force the federal government to begin planning for climate change. I had just joined the Department of Homeland Security (DHS), the third largest federal agency in the United States. Given the political noise that surrounded climate change then, no one immediately stepped forward to lead the DHS response. So, in the time-honored tradition of bureaucracies, the assignment to develop the department's adaptation plan fell to the newest arrival on the DHS leadership team: me.

The "no more" moment of the September 11, 2001, terrorist attacks spurred the U.S. Congress to revitalize security efforts. That trauma drove the largest federal government reorganization since World War II and led to the creation of DHS. In addition to its antiterrorism focus, DHS shoulders broad security responsibilities across its close to two dozen agencies, including emergency response through the Federal Emergency Management Administration (FEMA), immigration oversight through Immigration and Customs Enforcement (ICE), and protection of American waterways through the U.S. Coast Guard.

To respond to Obama's order, I assembled a task force. We members asked ourselves a basic question—one that in retrospect seems somewhat insubordinate to the president's direction. But until that question had a solid answer, I knew it would be difficult to garner enthusiasm for the task ahead. In the face of widespread skepticism about the reality of climate change, we asked ourselves, "Should the Department of Homeland Security in 2009, with all of its other responsibilities, care about the impacts of climate change?"

I also knew that answering this question would require the task force to consider the best available science about climate change. I am not a scientist. Nor were most of the other members of the task force. But based on my years as a judge presiding over thousands of legal proceedings, I knew that nonscientists make important decisions based on scientific evidence every day. Those decisions can even involve life and death when, for example, jurors rely on scientific evidence to weigh whether a criminal defendant deserves the death penalty.

Over several months, the task force heard from dozens of scientists, planners, and security experts, including the U.S. Navy, NASA, and the National Oceanic and Atmospheric Administration (NOAA). Participating in the task force gave me and others what I have come to consider the "climate aha moment," the moment when you begin to understand how profoundly climate change affects the systems humans have created and come to rely upon. In due time, as we learned about the projected hurricanes, wildfires, and droughts that could pummel America in the near future, the task force had its answer: DHS should care deeply about climate change. It would affect virtually all the department's missions.

Leaving stability behind

The evidence presented to the DHS task force showed that the long-lasting stability of the earth's climate had begun to shift. For thousands of years, humans had relied on past experience to predict how high the rivers would flow as the snows melted in the mountains, when the monsoon rains would come, and how hot it could get in the summer and how cold in the winter. As cities and states formed and populations exploded, recent history offered a reliable guide for when to plant crops and where to build wastewater treatment plants, lay train tracks, construct dams, build homes in river valleys, and create ports across the globe. In this stable climate, human civilization flourished.

But, starting around 1850, the Industrial Revolution set in motion future change. Even as the burning of coal, oil, and natural gas brought unprecedented prosperity and progress, it also spewed increasing amounts of greenhouse gases. The laws of physics dictate that higher levels of these gases in the atmosphere trap more of the earth's heat, thus causing planetary temperatures to rise and rendering the assumption of a stable climate untrue. The rate of GHG accumulation greatly exceeds anything seen in millions of years of geological history. Within 50 years, one to three billion people could "be left outside the climate conditions that have served humanity well over the past 6,000 years."[3] With stability waning, never-imagined, record-breaking events increasingly strain the human-built and natural systems upon which people rely.

To prepare for accelerating extremes, communities need to adapt. They should consider the future risk of climate change as they make decisions about how and where people should live. But imaginations continue to fail, at many levels. In 2014, the UN Intergovernmental Panel on Climate Change (IPCC), a group formed to assess the science related to climate change, concluded that even when adaptation measures were put in place, those measures often were designed to cope with relatively small variations in climate rather than with the larger, long-term changes caused by a warming planet. Similarly, a 2018 review of the corporate adaptation plans of over 1,600 of the world's largest companies revealed widespread reliance on the "enduring assumption" of a stable climate. This reliance predisposed corporations to cling to management approaches that were poorly matched to the anticipated scope and scale of climate change impacts.[4] That same predilection for assuming a stable climate has left flood regulations mostly reactive, responding to past disasters rather than trying to look ahead to future risks. In 2014, a group of retired U.S. military generals and admirals warned, "When it comes to thinking about the impacts of climate change, we must guard against a failure of imagination."[5]

The past is no longer a safe guide for the future

Let's look at Japan, a nation widely recognized as a model for disaster preparedness. Japan sits along the Ring of Fire, a massive horseshoe-shaped belt of volcanic and seismic activity coming from the floor of the Pacific Ocean. Close to 90 percent of the globe's earthquakes occur along the ring, and as a result, Japan has for centuries sought to reduce its risk. The country even gave the world the name for the giant waves that can follow earthquakes, tsunami. Since at least the Middle Ages, Japanese villagers have placed engraved stones on coastal hillsides high above the normal sea level to mark the height of past tsunamis, serving as warnings for future generations.

The small fishing town of Taro, located on the northeastern coast of the Japanese island of Honshu, is no stranger to the destructive force of tsunamis. In 1896, a tsunami struck the town, destroying every single building and killing over 80 percent of the residents. Almost five decades later, in 1933, another tsunami slammed into Taro, taking more lives and demolishing more buildings. In the wake of these events, the people of Taro vowed to build back better, or at least well enough to avoid similar destruction in the future. The community began constructing a huge seawall.

Over time, the seawall grew to dominate the entire town. The wall earned the nickname "the Great Wall of Japan," and the town itself assumed the name "Tsunami Boy." The structure became a source of civic pride, and in 2003, the townspeople officially commemorated it as a symbol of resilience and courage.

Then, at 2:46 p.m. on March 11, 2011, a magnitude-9.0 earthquake struck in the western Pacific Ocean. Its epicenter lay some 80 miles (129 kilometers) east of Japan, but its tremor shook Russia, Taiwan, and China. One of the most powerful earthquakes ever recorded, it spawned an enormous tsunami. Within minutes, the wave, which in some places measured as tall as a ten-story building, slammed into the Japanese coast, washing as far as six miles (9.7 kilometers) inland. The tsunami left an estimated 20,000 people dead or missing and destroyed over 120,000 buildings. It even washed away some of the ancient tsunami stone markers. Most famously, the flooding caused a nuclear disaster at the Fukushima Daiichi Nuclear Power Plant, depositing traces of radioactive material as far away as North America.

As the tsunami approached Taro, residents climbed atop the famous seawall. Racing toward shore, the wave stood more than five meters (16.4 feet) above the wall's tallest point. The water destroyed Taro's seawall, killing those who had sheltered there. The tsunami destroyed all of the seawalls, except for one, along the entire coast of the Tohoku region of Japan.

After this 2011 disaster, Japan redoubled its efforts to reduce the death and destruction from future disasters. Some towns prohibited house construction in vulnerable areas near the coast, relocating residents to higher ground. The Japanese government committed to spend more than $200 billion to build at least eight towns located safely above the reach of future tsunamis. The government also embarked on a plan to build stronger seawall defenses, including in Taro, designed to last at least 80 to 100 years.

Yet, researchers have identified a potentially fatal flaw in the government's plan. Japanese officials did not factor in sea-level rise when redesigning the failed Taro seawall. They asserted that the issue did not merit consideration because the seas were rising too slowly. In making that assumption, they neglected to consider one of the major impacts of climate change— accelerating sea-level rise. The Japanese government planners failed to "imagine" the risk that future tsunamis would cause even greater damage, despite a growing body of evidence that higher sea levels will make frequent small tsunamis taller and more deadly.[6]

The World Bank estimates that across the globe multi-billion-dollar infrastructure projects and master plans for tens of billions of dollars of investment frequently fail to consider the implications of future climate change. Already about 10 percent of the world's population lives in low-lying coastal areas vulnerable to flooding, a figure projected to increase between 5 percent and 13.6 percent more by 2100. In the United States, the world's largest economy, communities have clung to the belief that the climate is stable, focusing largely on preparing for risks like those posed by the current climate and recent past. As a result, American building standards and land-use practices do not yet reflect that the past is "no longer a sound basis for long-term planning."[7] The failure to imagine how bad it could get has added to the challenge of addressing climate change. Until planning for climate risk becomes routine, communities will continue to be vulnerable to greater damage.

Planning is essential

In ordering federal agencies to plan for climate change, President Obama sought to jumpstart national efforts to adapt through planning, a widely accepted method for driving greater preparedness. In response, our DHS task force created a climate change roadmap and implementation plan, while many other federal agencies also developed their first adaptation plans.

One of the best descriptions of the challenges of adaptation planning that I've heard comes from Sierra Woodruff, whom I first met when she came to work for me as an intern in the White House. Named by her father for the spine of mountains on the eastern edge of California, she developed a lasting interest in ecology after reading Jared Diamond's *Collapse* in middle school. At the White House, Woodruff took on the arduous task of sifting through the stacks of adaptation plans submitted by the various federal agencies, including the one DHS crafted. Later, she got her PhD at the University of North Carolina, focusing on adaptation planning and becoming a leader in the emerging field. As she describes it, the whole process requires combining many different academic disciplines and navigating rapidly changing conditions, akin to "building the plane while flying."

That's certainly how it felt to us at the department. The DHS task force had to plan even as the department contended with crises such as Hurricane Sandy, tens of thousands of Central American children arriving at the U.S. border, the Boston terrorist bombing, the spread of H1N1 influenza, natural disasters ranging from earthquakes to wildfires, and the Deepwater Horizon oil spill. Moreover, tackling the task is only the beginning. After reviewing hundreds of adaptation plans for cities and states within the United States, Woodruff has concluded that, so far, not much progress has been made on the "adaptation trajectory." Competing demands for resources and time, uncertainty as to what impacts will strike and when they will occur, siloed information, lack of a common metric to measure progress, and want of leadership—among other things—have impeded progress, not only in the United States but also across the globe. That's why planning efforts need to accelerate.

National governments play critical roles in spurring adaptation. Planning for climate risk can help drive acknowledgment of the risks of climate change; clarify how investments in hazard reduction can reduce overall damages; set priorities for public investments in adaptation and identify incentives for private investments; harmonize government programming; promote unified, "whole-of-society" as well as regional approaches to climate challenges; distribute resources and support for local decision-making; coordinate and bolster disaster response capabilities; direct funding to research; account for the needs of the most vulnerable; require adaptation of all government facilities and operations; and identify ways to measure progress.

Planning is particularly important when it comes to decisions about infrastructure and/or land use. The choice to build infrastructure can lock in development patterns for decades. A new road, for example, may lead to greater development in an area at higher flood risk, which, in turn, results in increased damages when extreme rainfall occurs. Proper high-level planning provides an opportunity to head off these poor decisions before communities commit. It also encourages decision makers to move beyond just considering how to climate-proof a particular project or structure to ensuring that systems and infrastructure sectors support climate resilience on a regional and national basis.

Engagement in nationwide planning efforts, which include all levels of government and cut across all sectors of society, could further drive transnational and cross-jurisdictional cooperation for climate threats that cross existing borders. National adaptation plans can achieve "a thorough, comprehensive new approach to physical, social, and economic planning."[8] One way to expand the impact of national plans is to accompany them with national laws requiring local communities to prepare their own plans. Research shows that cities in countries that have such legislation are five times more likely to have local adaptation plans.[9]

National adaptation planning, however, remains a work in progress.

When the 2015 Paris Agreement came into effect in 2016, it established a requirement for each nation, "as appropriate," to engage in adaptation planning processes, which could include "the process to formulate and implement national adaptation plans." By 2020, most member countries of the European Union had developed national adaptation plans, but according to the European Commission's evaluation of those plans, more work needed to be done to achieve adaptation goals. As of 2019, just 16 of the 153 developing countries had submitted completed plans to the

United Nations. A 2019 global assessment of "where government-led policy is happening, and where it is not, in coastal regions," concluded that although a high number of adaptation policies existed, national policy had not trickled down to regions or cities in the developing world.[10] Of 68 nations surveyed, almost a dozen had no national plan at all.

Because of this lackluster progress, in 2019, the Global Commission on Adaptation, chaired by Microsoft founder Bill Gates, former UN secretary-general Ban Ki-Moon, and International Monetary Fund head Kristalina Georgieva, judged adaptation efforts worldwide as "gravely insufficient."[11] The commission called for nations to create strategies based on the latest information to guide "where and how to build new communities and all other new investments in order to limit future exposure and risk."[12]

Creation of a national plan, however, does not guarantee that success will follow. Although some nations had pandemic plans that ran hundreds of pages, those plans did not stop the spread of COVID-19. Belgium reportedly did not even consult its plan during the pandemic, although global health officials had thoroughly reviewed it. In some cases, as in the United Kingdom, the plans relied on faulty assumptions, for example, that public health capabilities remained strong even after years of budget slashing or that during a crisis the government could purchase medical supplies "just in time" in the global economy.

National planning for climate change has revealed similar cracks. For example, Japan created its first national adaptation plan in 2015. The plan embraced a strategy of mainstreaming adaptation into government policy, yet as we saw, the Taro seawall's design failed to consider sea-level rise. Bangladesh created its

Climate Change Strategy and Action Plan in 2009, yet some non-governmental organization (NGO)-led resilience initiatives have stalled in the absence of longer-term government support, ending merely as pilot projects once resources are exhausted.

Planning for climate change requires an iterative process, one that gets refreshed regularly and builds muscle to support planning by local governments, the private and nonprofit sectors, and individual households. Steady, continued focus on the challenge will deepen understanding of climate change, allowing potential solutions time to marinate and improve. Engaging multiple stakeholders—including government agencies covering different sectors, ranging from agriculture to energy, and different levels of government, as well as private citizens—also increases the likelihood of success. As U.S. President Dwight Eisenhower, who had led the Allied Expeditionary Forces in World War II, once advised, "Plans are useless, but planning is everything."

What's at stake?

The United Nations predicts that in the next four decades, the world will double the amount of floor space in buildings worldwide. That's equivalent to adding a new New York City every month or adding the floor area of all of Japan's buildings every year. If the new construction is not designed to be resilient, populations across the globe will remain vulnerable to climate impacts.

When building occurs, getting investments in infrastructure right is vital. As the backbone of human civilization, infrastructure provides services essential to connecting people and promoting health, safety, and economic durability. When infrastructure fails, it causes cascading damage to society, disrupting commerce and undermining social welfare and security. Failure to adapt infrastructure to changing climatic conditions can leave whole communities and regions vulnerable (Figure 1.1).

All nations face challenges with their infrastructure. In developed nations, the infrastructure is often old, in need of replacement or repair, and not built to withstand the kind of climate impacts the planet is already experiencing. When worsening climate impacts hit existing infrastructure, it can fracture business operations, as Verizon Wireless learned during Superstorm Sandy in 2012.

Verizon has its headquarters in Lower Manhattan. A line on the wall of its building marks the height of flooding caused by Sandy. Fueled by climate change, Sandy brought a storm surge that overcame New York City's 12-foot barriers, barriers that failed to account

FIGURE I.I. In 2018 in Devon, United Kingdom, Storm Callum sent waves that overtook infrastructure like trains, roads, and homes.
Source: Theo Moye / Alamy Stock Photo.

for the foot (0.3 meters) of sea-level rise that had occurred in the area since 1900. Before the storm's arrival, Verizon had modeled the likelihood of more extreme events as a result of climate change. But the company had not appreciated its vulnerability to municipal infrastructure. Floodwaters poured into the company's facilities through municipal manholes, damaging two 90,000-cubic-foot vaults housing copper-based wiring systems. The saltwater corroded the copper, and Verizon's customers lost service. The company suffered a $1 billion economic loss and damage to its reputation (Figure 1.2).

In addition to communications systems like Verizon's, extreme events have battered electrical utilities in recent years. In the United States, the electric grid has failed to keep pace with new climate extremes. In California, the threat of sparking wildfires has driven electric companies to preemptively shut off power supplied

FIGURE 1.2. Hoses drain millions of gallons of water out of the Verizon Headquarters in New York City after record storm surges brought by Hurricane Sandy submerged several floors in 2012.
Source: REUTERS / Alamy Stock Photo.

to communities. At the same time, droughts have reduced hydropower in the Northwest. In 2021, a blast of Arctic air caused over 4 million people in Texas to lose electricity because power companies could not operate in conditions of extreme cold.

China faces challenges due to its heavy investment in dams decades ago. Now, as a result of more punishing levels of precipitation, many of its 98,000 dams—the most of any country in the world—are in poor condition and subject to extreme water levels that threaten their integrity.

Developing nations face even greater challenges as they experience rapid rates of urbanization and population growth. Just consider the world's megacities. Most of them are located within low-elevation coastal zones, many of which are in Asia. Coastal megacities have millions of people crammed into densely populated informal settlements, with more people arriving every day. These cities face the dual threat of sea-level rise and greater storm surge, and probably more intense extremes of heat as well. Many small- and medium-sized cities face similar threats.

A recent study concluded that storm surges are now eight inches (20.3 centimeters) higher than they were in 1900.[13] This means that floodwaters move ever further inland, eroding more beaches, destroying more infrastructure, causing the loss of more homes, and forcing more people to move. Many cities lack adequate structural protections to avoid either the initial trigger event—like storm surges—or the cascading breakdowns of basic services that follow. To prepare, cities will need help with understanding their risks and what to do about them. If they overlook climate risk as they make choices about where people live and what infrastructure to build, the investments leaders make today may wash away in the next storm.

One way to get information is by modeling possible futures. Climate modeling starts with global models since the climate system itself is global. The global climate models used to inform climate projections are at a high resolution of 500 kilometers (310 miles), which makes them too imprecise to inform most planning decisions on the ground. As a result, communities and businesses have issued a clarion call for "down-scaled" data to better inform choices going forward. A wide variety of methods can filter coarse-resolution data from global models into more useful products, but lack of widely available down-scaled data means that growing risks may go undetected. A 2020 study by Climate Central, an American nonprofit organization dedicated to communicating climate science, revealed how alarmingly large these risks have grown.[14]

Climate Central embarked on a project to determine the accuracy of sea-level projections currently available to communities in less developed parts of the world. Climate Central scientists knew that sea-level-rise models have tended to focus on how much seas will rise. The scientists decided to focus on the other side of the equation—the elevation of the land. How much damage sea-level rise will inflict depends on whether a town or a building is three feet (0.9 meters) above the water's edge or ten feet (three meters). Projecting sea-level rise requires not only estimating the level of rise but also comparing it to existing land elevations.

Although Europe, the United States, and Australia have access to sophisticated elevation information through LIDAR, a state-of-the-art remote sensing technology, other places in the world do not, notably South Asia. Where LIDAR is not available, researchers have relied on a data set that has an average error rate of six feet. Climate Central decided to conduct its own analysis to see if it could reduce the uncertainties. Using machine learning, Climate

Central combined existing data sets to reassess elevation, or in its words, to "get a better view of the ground beneath our feet."[15]

The improved data indicated that sea-level rise threatens far more land and people than previously thought. Earlier estimates had identified 79 million people living on land that would flood every year by 2050. The new data showed more than three times that number, 300 million people, facing flooding every year. The number of people living below the high-tide line, and therefore subject to daily flooding by 2050, jumped from 38 million to 150 million. Wide data gaps exist in developed nations as well.

Additionally, global climate models are still not well-suited to predict extreme events. Researchers studying the Black Summer bushfires of 2019–2020 determined that the global climate model information produced by the Coupled Model Intercomparison Project, an effort that coordinates modeling results across dozens of centers for the IPCC reports, missed the fires that raced across southeastern Australia.[16]

To achieve the necessary revolution in understanding that governments, communities, businesses, and individuals require to adapt successfully, they must have access to locally relevant climate information and data. In other words, they need access to "climate services," the products and services that translate climate science to aid decision-making. Climate services help make climate risk real.

Making risk real

Many researchers have identified the need to support decision makers with tools, modeling, scenarios, and other services— ideally codeveloped with and tailored to the needs of end-users—to make informed choices about climate risk, including both mitigation and adaptation measures. With growing climate risk, the demand for such services from all sectors will only increase. Sadly, access to climate services differs dramatically between the haves and the have-nots.

In 2009, the World Meteorological Organization developed the Global Framework for Climate Services (GFCS) to improve "incorporation of science-based climate information into planning, policy, and practice."[17] As of 2020, only four countries, Switzerland, China, Germany, and the United Kingdom, provided advanced climate services according to the GFCS. A 2019 study concluded that many countries across the globe lacked the capability and capacity to develop and deliver climate services.[18] About one-half of the world's nations lack effective climate services systems. Africa and the small island developing nations lag far, far behind.

Meanwhile, the entry of private companies into the field has ignited, in the words of Tulane University Professor Jesse Keenan, "a climate intelligence arms race," to serve those who have the means to pay.[19] This arms race has resulted in the development of a growing array of mostly privately developed products and services designed to inform efforts to prepare for climate change. One of the most long-standing tools is catastrophe modeling, which can help decision makers project damage from catastrophic risk.

Catastrophe modelers got their commercial start after Hurricane Andrew, a Category 5 storm, flattened neighborhoods across the U.S. state of Florida in 1992. Before the storm hit, property insurers and reinsurers had solely relied on past claims to estimate insurance risks. The storm caught the industry by surprise when it caused unprecedented losses and insurance claims totaling $15.5 billion. Eight insurance companies went bankrupt in the aftermath. Before this, one insurance veteran had estimated that a storm the size of Andrew would cause only $4 to $5 billion in damages.

In the face of financial ruin, insurance companies began hiking rates, canceling policies, and threatening to stop doing business in the state altogether. The Florida legislature stepped in to buy time to find a solution to the crisis. Eventually, the companies collaborated with state insurance regulators to permit the use of catastrophe models to assess risks from natural hazards. An industry was born.

Nancy Watkins heads an actuarial consulting practice in San Francisco, California, at a company called Milliman, which analyzes the financial consequences of risk, including the growing risks from climate change. She believes that catastrophe models have an important role to play in avoiding failures of imagination. By running thousands of simulations, catastrophe models can project a range of outcomes for high-consequence events that occur with low frequency.

Watkins has analyzed the insurance industry's failure to imagine the back-to-back wildfire seasons of 2017 and 2018 in the state of California. California, which is the largest insurance market in the United States and the fourth largest in the world, prohibits the use of future-focused models for insurance pricing. Instead, the state requires insurers to price risk based on the past. Insurance losses from those two fire seasons wiped out double the combined underwriting profits of the previous 26 years, sticking insurers with an aggregate underwriting loss of more than $10 billion. Watkins calls the unprecedented losses from wildfire in just those two years a "wake-up call" for the industry (Figure 1.3).

FIGURE 1.3. A neighborhood burned to the ground, a pinprick out of the 6,700 buildings lost to the 2018 Camp Fire that swept through Paradise, California.

Source: © Carolyn Cole 2021. Los Angeles Times. Used with Permission.

Increased wildfire risk is also occurring around the globe, including in the Mediterranean region, Mexico, Brazil, Korea, and parts of Australia, China, and sub-Saharan Africa. As a result of the rapidly changing risk picture, Munich Re, one of the largest reinsurers in the world, warned insurers that following California's approach of using a long claims history to price insurance will "seriously underestimate the risk."

Since Hurricane Andrew, catastrophe modeling has greatly expanded, moving into areas beyond earthquakes and floods to cover threats like pandemics. As interest in climate risk has grown, the modeling industry has moved to the "front line" of assessing climate risk for corporations and governments.[20]

Catastrophe models are much finer resolution than the global climate models, reaching as low as ten-meter resolution. The models, however, typically only examine shorter-term time horizons of up to five years. They are particularly useful for modeling extreme events and calculating financial impact. It allows decision makers to examine possible future outcomes for hazards like assessing the viability of elevating homes to adapt to climate-worsened flooding.

But models are not a panacea for making the right choices about risk reduction. They have limitations and will only be useful when those limitations are understood and accounted for. One potential problem, for example, is that people can mistake the model for reality. As Watkins put it, "People tend to believe that all that is being measured is all that exists." They may tend to believe an incorrect model even if other facts indicate danger ahead, something akin to relying exclusively on Google Maps directions and then finding yourself driving into a lake. When facing the rising infection rates of COVID-19 in the spring of 2020, the U.K. prime minister was so confident in the forecasts from British modelers that he delayed locking down the country until two weeks after British emergency rooms had become overwhelmed.[21]

Models may also fall short in capturing human behavior and the consequences accurately. For example, pandemic modeling may show how many people get sick and die, but not what happens to the economy when people stay home because they are afraid. Models also struggle to capture the human role in wildfire ignitions and how well suppression efforts will keep blazes from spreading. In other words, the quality of the outputs from the models depends on the underlying assumptions and the level of analysis. With climate change, those assumptions have a thick extra layer of complexity since, in the words of one insurance expert, "It is harder to model a moving target."

As the world learned with COVID-19, each model may tell a different story. Echoing Dwight Eisenhower's observation on planning, the statistician George Box quipped, "All models are wrong, some are useful."[22] In other words, as Watkins's firm concludes, "a good model can provide users with significant value in spite of outstanding uncertainties as to model precision."[23]

In addition to catastrophe modeling, other types of climate services have experienced exponential growth. These services include developing precise hazard data for critical infrastructure assets as well as improved forecasting, both of which are highly cost-effective; expanded use of remote sensing, in combination with artificial intelligence and machine learning, to enhance vulnerability assessment; investments in data visualization systems to improve understanding of risk; greater use of scenarios to analyze possible future conditions; and the deployment of satellite imagery to identify emerging physical risk.

But the cost of such services runs high. They require the expertise of a panoply of professionals, including meteorologists, statisticians, computer scientists, engineers, and actuaries. Given the complexities involved, creation of locally tailored climate risk assessments and analysis is beyond the capabilities of all but the largest businesses and well-resourced governments. That means that many communities, small businesses, and households could lack sound risk information.

With growing climate risk, the question then becomes who will pay? To avoid leaving communities behind, NGOs, philanthropies, and others have begun to underwrite a certain level of climate services. For example, reinsurance companies have developed the Oasis Loss Modelling, an open-source platform for catastrophe modeling, while European researchers created CLIMADA, a global multihazard decision support tool for assessing economic risk. Ensuring broad access to climate services, like these and others, is the linchpin to averting more failures of imagination.

Leadership matters

In the summer of 2009, I watched, horrified, as my daughter collapsed unconscious in front of our doctor. The diagnosis? She had H1N1 influenza. We left the doctor's office with me clutching her in one arm and a prescription for the antiviral medication Tamiflu in the other.

Her illness came during a virulent outbreak of H1N1 influenza that first hit the United States in the spring of 2009. This dangerous flu strain would eventually cause the hospitalization of over a quarter-million Americans and the deaths of more than 12,000. With the public scared, Tamiflu ran short. On the day my daughter fainted, I raced from pharmacy to pharmacy in search of the promised elixir. I never did get it, but thankfully, my daughter recovered anyway.

As infections continued to spread later that year, Secretary of Homeland Security Janet Napolitano asked me to run a task force at DHS focused on preparing for biological threats. The DHS Biological Leadership Group aimed to ensure that the department was as ready as possible for biological threats, from an influenza pandemic to an aerosolized anthrax attack. After my own family's frightening experience with H1N1, I needed no prompting to get started right away.

The DHS Biological Leadership Group met regularly for the four years I spent at DHS. We looked at how DHS could best coordinate with state and local governments, including with their public health departments. That examination unveiled how public health departments across the nation hungered for funding and how diagnostic capabilities remained woefully inadequate. We looked at security issues surrounding research laboratories and how to address the growing risk of zoonotic diseases, the diseases that spread from animals to humans, as people have increasing contact with animals in the wild. We also updated the department's pandemic plans.

We had the advantage of building on a strong foundation. In the United States, national planning for pandemics had started in earnest after a fiasco in response to a flu outbreak in 1976. In January of that year, swine flu flared at an army base, Fort Dix in New Jersey, quickly sweeping through the 19,000 personnel at the base. Medical experts identified the disease as a new strain of swine influenza. Fearing a widespread pandemic, U.S. President Gerald Ford vowed that every man, woman, and child would receive a vaccination. Within two months' time, 45 million Americans received shots in the arm. Reports soon emerged of a connection between the vaccine and a serious condition called Guillain-Barre syndrome. The vaccination campaign ended abruptly. Like clockwork, federal officials ordered an investigation into the "swine flu affair," spurring increased attention to pandemic preparedness.[24]

The incident prompted officials to embrace nationwide planning, releasing the first national pandemic plan two years later. Public health preparedness efforts redoubled after the anthrax attacks just one week after the 9/11 terrorist attacks. When avian flu spread through Asia in 2004, the U.S. government poured money into state and local preparedness efforts and procurement of vaccines and antiviral drugs. By the end of 2005, the nation had a new National Strategy for Pandemic Influenza, and by 2006, it had created both a government research arm focused on development of medical countermeasures and a mechanism to assess state pandemic planning efforts. The Ebola crisis from 2014 to 2016 drove focus further on disease containment. By the end of Obama's presidency, several new pieces had been added to the pandemic preparedness puzzle.

One puzzle piece was the creation of the Office of Global Health within the National Security Council (NSC) in 2016. In late 2015, I had gone to the West Wing office of then-Vice President Joe Biden to talk about biodefense. The vice president wanted to know where responsibility for preparedness to biological threats should fall within the U.S. government and whether responsibility should move to the Office of the Vice President as some outside experts had advocated. Seated around his fireplace, my colleagues and I advocated that it made little sense to require the vice president's office to assume the mantle for biodefense, given the urgent demands on its attention. Rather, we recommended that biodefense should remain within the NSC. The wisdom of this approach surfaced when the Office of Vice President Mike Pence assumed the lead for the COVID-19 pandemic response and quickly fumbled.

Another piece the Obama administration added to the puzzle was an updated, 69-page pandemic "playbook" designed to guide the new administration.[25] Along with a directory of available government resources to combat the spread of infectious disease, the

Playbook for Early Responses to High-Consequence Emerging Infectious Disease Threats and Biological Incidents warned that the "American public will look to the U.S. government for action." To hammer home the need for readiness, Obama officials hosted a simulated exercise focused on the emergence of a novel influenza strain for the incoming administration. Originating in Asia, the imagined disease swept around the globe, causing health care systems to crash and threatening to rival the 1918 Spanish flu epidemic.

All the planning in the world does not matter, however, if it is ignored. Beginning in 2017, the Trump administration changed course. It closed the global health preparedness office at the NSC. In 2019, it quietly removed the last American doctor working inside the Chinese Centers for Disease Control and Prevention (CDC), shuttered the Beijing office of the American CDC, and suspended a U.S.-sponsored program in China that worked with a laboratory in the city of Wuhan to detect the spread of disease from animals to humans. Meanwhile, concern about the nation's level of preparedness intensified among public health experts.

Tom Inglesby is one of the world's foremost experts on biological threats. As director of the Center for Health Security at the Johns Hopkins University in Maryland, he organized a tabletop exercise in 2018 to test preparedness for a global pandemic. The simulation dropped participants, including policymakers, health officials, and business leaders, into the middle of an uncontrolled outbreak of a coronavirus from South America. The fictional disease named CAPS had crippled trade and travel and thrown the global economy into a tailspin. As the exercise progressed, it projected citizens rioting in the streets trying to get their hands on a fictional antiviral, while researchers worked feverishly to develop a vaccine.

The outcome of the simulation was not reassuring. Inglesby warned that the nation remained ill-prepared and concluded that "progress is possible only with effective leadership."[26]

COVID-19 emerged in South Korea and the United States on the very same day in January 2020. President Trump downplayed the significance of the arrival of the virus, insisting, "It's one person coming in from China. We have it under control."[27] By late February, South Korea had the greatest number of diagnosed cases of any country except China. By late May, however, South Korea reported fewer than 300 deaths, while the number of fatalities in the United States soared to over 100,000. The reaction of the leaders of the two countries could not have differed more.

Once the pandemic began to spread rapidly in the United States, President Trump veered away from the pandemic response strategies that federal, state, and local officials had relied upon. Instead, the White House put competing groups in charge, one headed by Vice President Mike Pence and another by the president's son-in-law, Jared Kushner, both struggling for presidential attention. Kushner's group concocted a secret plan for testing for the virus. The plan, however, never materialized. It "just went poof into thin air," according to one participant.[28] Competing lines of effort disconnected the systems that public health authorities had spent years developing to handle health emergencies.

One *Washington Post* story detailing the government's pandemic response carried the headline "70 Days of Denial, Delays and Dysfunction."[29] In early March 2020, a former colleague of mine on the DHS Biological Leadership Group texted me. He had devoted his career to ensuring the federal government's readiness for biological threats, including pandemics. "What are they waiting for?" he asked. The United States, meanwhile, continued to slumber. Inaction violates a basic principle about pandemics and climate change.

Early action is essential. The South Korean government understood this in its pandemic response. After a flawed response to MERS in 2015, South Korea had invested in reforms to boost preparedness. As a result, the nation could immediately mobilize like "an army."[30] By mid-March, South Korea had built the widest and best-organized testing program in the world up to that time. The country had tested more than 270,000 people, or more than 5,200 tests per million residents. It even pioneered the use of drive-through testing inspired by Starbucks. The United States, in contrast, had done only 74 tests per million. By the end of that month, the number of new South Korean cases had dropped to just 100 per day compared to 900 per day in late February in the United States. On May 5, the United States had more than 19,000 new cases, while South Korea, a country of 50 million people, only had three. Its success with handling COVID-19 allowed South Korea's economy to expand beyond its pre-pandemic level by April 2021.

In the words of *The Economist*, the economic pressure to reopen and the White House inaction "scuppered any chance America had to produce a national strategy."[31]

Leadership also matters in the boardroom. Unfortunately, far too few companies, at least before COVID-19 hit, had moved dealing with strategic risk to the top of the board agenda. When the virus struck, few companies had enough supplies or redundancy in supply chains to cope with the disruptions. The "hyper-efficient, just-in-time, low-cost" global supply chain revealed its vulnerabilities, according to the international accounting firm KPMG. KPMG warned that COVID-19 was "not a one-off" and urged companies to incorporate risk management into strategic planning for the growing extremes ahead and to "routinely simulate" how such events can disrupt their operations.[32]

The story of COVID-19 shows how much leadership matters. But it may matter even more when it comes to preparing for the impacts of climate change. FM Global, a global insurance company with a focus on large corporations, found that more than three-quarters of CEOs and CFOs at the world's largest corporations in 2020 admitted that they were not prepared for the adverse impacts of climate change.[33] Because of the long-lasting changes that climate change brings, it requires sustained leadership well into the future. It requires making sure that governments and the private and nonprofit sectors remain focused on cutting greenhouse gas emissions as they build resilience to the harmful impacts. It requires developing a national adaptation plan informed by risk modeling that integrates resilience considerations into decision-making. It means providing meaningful risk information to decision makers and screening investments to make sure they can not only hold up to the ravages of future climate extremes, but also bolster long-term resilience.

To repeat, leadership matters. After the Japanese bombed the U.S. naval base at Pearl Harbor in 1941, the U.S. Congress sought to identify what had gone so horribly wrong. It found failures to centralize authority and clearly allocate responsibility for operational, financial, managerial, and executive work. When investigators examine the causes of the United States' stumbles in the face of COVID-19, a lack of leadership will also surface.

No matter how sophisticated the planning, the mapping, or the modeling, the final test will come down to leadership, whether in combating a biological threat or addressing climate change. As one longtime preparedness expert opined, "If a president chooses to ignore advice, it doesn't really matter how you're organized."[34]

CHAPTER 2

PREPARE FOR CONCURRENT, CONSECUTIVE, AND COMPOUNDING DISASTERS

Sediakan payung sebelum hujan
Prepare the umbrella before it rains

—*Malay proverb*

Before COVID-19 began its race around the world, experts had warned that for pandemics, it was a question of when, not if. Recent near misses—severe acute respiratory syndrome (SARS) in the early 2000s, H5N1 (bird flu) and H1N1 (swine flu) in the late 2000s, Ebola in 2013, and then Middle East respiratory syndrome (MERS) in 2015—as well as evolving research, had convinced scientists that a global pandemic was a virtual certainty. At the Department of Homeland Security (DHS) Biological Leadership Group, we treated the eventual emergence of a previously unknown pathogen as a given. We had a sense of how the spread of disease might unfold, likely jumping from animals to humans. We also suspected that its spread might go undetected at first. But once the disease found more human hosts, it would probably explode into the population in the absence of known treatments or cures. It would, we feared, stress poorly funded public health systems to the breaking point. What we didn't know was the timing. Our preparedness efforts for biological threats continued, but without the benefit of knowing when a pandemic would occur—next year or in one, two, or three decades' time?

When COVID-19 did emerge, it unfolded largely along the lines our DHS Biological Leadership Group had imagined. Tragically, the contagion exposed a lack of preparedness among many governments and businesses. The virus disrupted supply chains, jumped borders, and crushed health care systems. What we had failed to imagine, however, is that, even as a contagious disease sows death and economic destruction, climate-fueled extremes could drive people out of their homes into the path of the disease. As COVID-19 unfurled, flooding covered more than a quarter of the land in densely populated Bangladesh, thirty named storms formed in the Atlantic basin, and Californians started using a new word—gigafire—to describe a single wildfire that scorches over a million acres.

When multiple disasters occur within a geographic region or close in time, their impacts can aggregate, undermining efforts by both people and their governments to cope. A rapid sequence of relatively modest weather events, or a cluster of them, may cause more damage than a single monstrous event. Climate change increases the risk that concurrent or sequential disasters will bring communities and even countries to their knees.

Current approaches for disaster response require an overhaul to address the burgeoning threats from climate change impacts. Four crucial steps can help countries prepare for a future of concurrent, consecutive, and compounding disasters. First, acting before disaster strikes lessens damage and allows communities to bounce back faster. Second, protecting supply chains avoids panic and economic loss. Third, stockpiling acts as an extra level of insurance when compounding events occur. Fourth, strengthening emergency response capabilities means that aid arrives earlier and leads to a speedier recovery. In addition, treating every disaster as a "no more" moment drives continuous improvement in disaster recovery capabilities. Taking these steps offers better protection for communities, including thwarting criminality that may take root when disaster response fails.

Step one: Acting before disaster strikes

Acting early can save lives and money. According to the United Nations, every $1 spent on disaster preparedness can save $7 in damages. Communities, businesses, and households should prepare for calamity before it hits. The stakes are high, especially for the low- and middle-income countries, where 92 percent of the deaths attributed to internationally reported disasters have occurred since 1990. The global community should prioritize providing early warnings before disaster strikes as well as funding to prepare for predicted disasters through "forecast-based action," which can inject cash into a household before a storm makes landfall, for example. Both approaches allow people to protect themselves and their assets ahead of calamity, but both also require access to accurate weather forecasting, something that still remains elusive in many parts of the globe. Weather prediction services exist at the national, regional, and global levels, but large sections of the world struggle to obtain understandable and usable forecasts.

A lack of readily available, tailored weather forecasts has meant that extreme events can catch people unaware and unprepared—sometimes in their sleep. That's what happened in Somalia in 2019, when the East African country experienced the worst flooding in its recent history and unexpected floodwaters poured into homes as families slept. The deluge covered 85 percent of the town of Beledweyne with its population of 400,000 residents. Somalian President Mohamed Abdullahi Mohamed called the destruction "beyond our capacity," as he pleaded for aid.[1] Not long after, the UN Development Programme launched a $10 million initiative to, among other things, expand weather monitoring to help prevent future surprises.

To produce accurate forecasts, meteorologists need precise weather measurements about local conditions, such as humidity, precipitation, and barometric pressure, as well as weather models that then use those measurements to inform local forecasts. Africa lacks both tools. A large percentage of its weather stations do not yet report accurate data, and without accurate data to inform them, local weather models are hard pressed to make more accurate forecasts, if they were available. "Garbage in, garbage out," is how Antonio Queface, an atmospheric scientist in Mozambique, described the problem.[2] Not surprisingly, Africa suffers from the world's least developed weather network, and what does exist is in a state of deterioration.

The dearth of reliable weather data has led to greater use of another forecasting method borrowed from developed countries, weather satellite information. In 2019, the British government funded an initiative to provide real-time satellite data on storms to meteorologists in Africa online. But that program has encountered its own hurdles, ranging from unreliable internet connections to insufficient training of meteorologists to interpret the data properly, and not enough distribution channels for sharing the forecasts. Moreover, the presence of clouds can interrupt the ability of satellites to make optical observations.

Better data, monitoring, and modeling infrastructure are only part of the solution, however. The science of meteorology has historically assumed that the climate is stable, which is clearly no longer true. A 2019 study by Stockholm University in Sweden found that ongoing climate changes make it more difficult to predict certain aspects of weather, including summer downpours, which could affect the ability to prepare for flooding.[3] As the planet continues to warm, researchers will need to improve both the science and the practice of meteorology to ensure that future forecasts are as accurate as possible. Doing so will prove particularly important when it comes to the use of early warning systems and forecast-based action to build disaster preparedness.

Warning saves lives

As we saw in Chapter 1, when the U.S. Centers for Disease Control and Prevention flagged a suspected disease threat from Wuhan, China, on January 2, 2020, U.S. government leaders neglected to follow up. When leaders lead, however, early warning systems can save lives. Bangladesh, one of the most disaster-prone countries in the world, has proven it.

In 1970, Cyclone Bhola, the deadliest tropical cyclone ever recorded, slammed into what is now Bangladesh. With a death toll of half a million people, the storm propelled the country to invest deeply in early warning systems. After Cyclone Sidr took 3,000 lives in 2007, the government renewed its commitment to cyclone readiness. State-of-the-art meteorological radar stations now provide minute-by-minute weather updates that identify approaching storms long before land-fall. Bangladesh has also created thousands of cyclone shelters, many designed to serve as schools during ordinary times. The payoff was evident when Cyclone Amphan struck in May 2020.

Cyclone Amphan had intensified as it crossed unusually warm waters in the Indian Ocean, eventually becoming the strongest storm ever recorded in the Bay of Bengal up to that time. As Amphan approached, Bangladeshi workers delivered pandemic supplies, such as masks, sanitizers, soap, and hand-washing sta-tions, to cyclone shelters. The Bangladesh Red Crescent Society mobilized 70,000 volunteers to issue warnings in coastal areas. Teams went door to door to warn residents about the approaching danger. Volunteers with megaphones deployed on foot, bicycle, rickshaw, and motorbike, urging evacuation (Figure 2.1). Text

FIGURE 2.1. Thousands of volunteers warn citizens in Bangladesh to evac-uate before Cyclone Amphan makes landfall in 2020.
Source: Photo by KAZI SHANTO/AFP via Getty Images.

messages alerted people to the incoming storm. Before the cyclone made landfall, two million people and 500,000 livestock were evacuated. Thanks to Bangladesh's early action, Amphan caused fewer than two dozen deaths.

As Bangladesh's experience has proven, early warning systems save lives and livelihoods. Despite their impressive track record, deployment of these systems has lagged globally. The UN Framework for Disaster Risk Reduction, adopted by 187 nations in 2015, specifically called for greater deployment, and the World Bank has estimated that in developing countries a $1 billion investment in early warnings could provide benefits ranging from $4 billion up to $36 billion.[4] Despite the compelling case for early warning systems, multiple barriers have slowed implementation.

For example, the value of the systems may vary depending on the lead time or the specificity of the predictions. If the lead time is short, the system may not provide people enough time to act. On the other hand, if the predictions lack specificity or are overstated, people can become complacent, ignoring future calls for evacuation even though the danger is real. Some cities, including many in Asia, lack the necessary emergency management infrastructure to support early warnings, and without that infrastructure warnings can prove useless. Further complicating matters is that the cost-benefit analysis of any particular system poses thorny questions, like how much benefit does a system provide when it warns against a low-frequency event that carries uncertainty as to timing and scope, or what value is appropriate for quantifying the avoidance of loss of life.

Given the growing risks from climate change, threatened populations can wait no longer. Authorities need to overcome barriers to ensure that people receive warning of impending disaster in time to save lives. Developing, deploying, and improving early warning systems is essential to building resilience to accelerating climate impacts.

Not waiting for disaster

In addition to early warnings, authorities should also promote forecast-based action, a promising new type of early humanitarian aid. Forecast-based action changes the centuries-old paradigm of delivering assistance after disaster strikes. Instead, it focuses on helping people before calamity hits by supplying them with materials and/or funds in advance. But like early warning systems, forecast-based action's success depends on sound weather forecasts.

In Africa in 2015, Uganda became the first country to use forecast-based action. Ugandan officials had received a forecast for heavy rains in a region about 300 kilometers northeast of the capital city, Kampala. Flooding had previously surprised residents in that area in 2007, and in 2015, they were suffering from an outbreak of dysentery. Based on the forecast, which the Uganda Meteorological Agency verified, the Uganda Red Cross distributed supplies before the rains arrived. Residents received flood preparedness items, fuel cans, bars of soap, and a month's supply of water purification tablets to prevent further spread of dysentery and other waterborne diseases. Even though, as predicted, the area flooded heavily, no emergency services were needed thanks to the preparations made in advance.

The Ugandan experiment made history. For the first time in its 150-plus years of existence, the International Red Cross provided humanitarian assistance in advance of a disaster based on a flooding forecast. "With such a timely disbursal," said the secretary general of the Uganda Red Cross, "we hope to avoid potential catastrophe before it even happens, supporting people to continue working and going to school."[5]

From that start, forecast-based action programs have sprouted in other locations. Under these programs, meteorological services, humanitarian aid providers, and other stakeholders work together to determine in advance what kind of forecasts (for example, a tropical cyclone making landfall within 30 hours) and what level of event (for example, anticipated wind speeds exceeding 150 kilometers per hour) will trigger the release of funds or supplies (for example, seeds or animal-care kits or provisions of cash payments). Then, when such a forecast occurs, the organizations distribute the materials and funds using a pre-established protocol.

Forecast-based programs have the potential to "revolutionize disaster risk management" by explicitly linking early warnings to early action.[6] Indeed, even during the pandemic, the World Food Programme (WFP), a branch of the United Nations and the world's largest humanitarian organization focused on food security, managed to provide cash to over 30,000 people threatened by monsoon flooding in 2020. At-risk households used the money to purchase necessary items and evacuate themselves and their livestock before the floods began. Moreover, studies so far have reported positive financial results with the programs, with research predicting a return on investment of $34 for every $1 spent.[7]

But as with early warning systems, scaling up forecast-based action programs faces hurdles. Without access to accurate forecast data, forecast-based programs risk depleting funds when no crisis in fact materializes. The programs can require substantial upfront funding to conduct vulnerability assessments and register households. Government officials may lack the capacity or the interest to engage in the necessary planning to determine triggers and distribution mechanisms. Corruption remains a risk as well. Scaling up will require developing the necessary expertise within both governments and humanitarian organizations to cover more hazards, more locations, more people, and a greater range of responses. Given its great promise, however, forecast-based action seems primed to flourish.

Step two: Fattening supply chains

Before COVID-19 began its spread, few outside the disaster management community or the medical and public health professions gave much thought to access to personal protective equipment (PPE). Nor did the number of ventilators in a given community merit discussion among friends. As billions began to fear for their own health and that of their loved ones, the availability of PPE became of great interest to ordinary people. And the more they learned, the more they realized that supplies were hard to come by.

Even in the wealthiest nation in the world, people could not find basic purchases, such as hand sanitizer, protective masks, disinfectant, and toilet paper. The United States had depended on supplies of medical equipment from other countries, which, once the pandemic came into view, closed off shipments to take care of their own citizens. In the spring of 2020, China, the world's largest manufacturer and maker of over 40 percent of the world's masks, gloves, goggles, visors, and medical garments, established new export restrictions. Those restrictions quickly convulsed the supply chains for critical medical products. When India could not import the Chinese chemicals it used to manufacture generic drugs, Indian suppliers stopped exporting acetaminophen and antibiotics. As global trade shriveled, countries were left scrambling to locate new suppliers.

People around the world discovered that widespread reliance on "just-in-time" delivery of medical items by companies and governments left them vulnerable (Figure 2.2). With the demand for critical materials surging, the U.S. government had to focus on "managing shortages" rather than "managing resources," Pete Gaynor, administrator of the Federal Emergency Management Administration (FEMA), told the U.S. Congress.[8] Dependence on

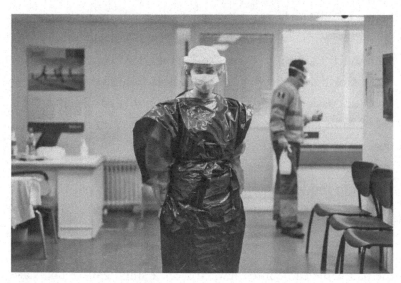

FIGURE 2.2. An emergency room nurse improvises personal protective equipment (PPE), sporting trash bags and a privately lent face shield in 2020 in Huesca, Spain.

Source: Photo by Álvaro Calvo / Getty Images.

overseas manufacturing had become "unsustainable," according to French President Emmanuel Macron.[9]

Other shortages unrelated to health care cropped up as well. For example, the spread of COVID-19 affected production in factories along the Mexico-U.S. border. In April 2020, 13 workers at a car-seat manufacturer died, as did at least three workers at an electrical components company. The Mexican government subsequently ordered the factories closed, threatening legal action against those that remained open. Similar bottlenecks occurred when air freight capacity dropped as scheduled flights ground to a virtual halt and ships were placed in quarantine or could not travel freely to pick up goods. In March 2020, a survey of Fortune 1000 companies showed that 950 of them had suffered supply chain disruptions. The supply chain woes exposed by the pandemic are a sobering harbinger of the problems that lie ahead on a warming planet. With more frequent, concurrent disasters fueled by climate change, supply chains will face even greater threats.

Global supply chains have existed for thousands of years, likely going back to prehistoric times. The ancient Greeks built cargo ships to service trading routes that extended throughout the Mediterranean region. The Romans traded with the people of Kerala, India, to obtain pepper in the days of Augustus. Since the 1990s, supply chains have stretched around the planet to every corner of the world, growing global trade by 350 percent—twice as much as the increase in global economic output. This "golden age" of trade has brought greater efficiency, lower costs, easier procurement, and an explosion in consumer goods.

The global supply chain lets companies keep their inventories lean, relying on just-in-time delivery. Multinational corporations can take advantage of cheap labor, global communications, and technology capabilities to spread manufacturing across the planet. New supply chains allow for the shipment of raw materials to manufacture in bits and pieces. Materials and components routinely cross multiple international borders before being finished into a final product. The thinner and longer the chains become, however, the more they grow in complexity, thereby increasing points of vulnerability.

Extreme events can break thin chains, causing widespread disruption. In 2021, high winds and a dust storm blew a 400-meter-long container ship sideways in the Suez Canal, one of the world's busiest shipping channels for oil and other products. With the vessel blocking traffic both ways for six days, losses to global trade mounted to over $9 billion per day. Climate change impacts like sea-level rise and worsening storms will increasingly bedevil the free flow of goods.

Consider the case of Puerto Rico. Before Hurricanes Maria and Irma, the U.S. territory and Caribbean island, home to about four million people, had developed into a global hub for the manufacture of pharmaceuticals and certain medical supplies. With some 40 manufacturing sites, the island is the source for 13 drugs that have no other supplier. It is also home to one of the leading manufacturers for most of the United States' supply of sterile saline solutions, which are used in treating wounds and dehydration. The year before the storms struck, $14.5 billion worth of pharmaceutical supplies manufactured in Puerto Rico accounted for 70 percent of the island's exports.

In the wake of the two back-to-back hurricanes, manufacturing capacity plummeted. The pharmaceutical and supply companies activated business continuity plans and deployed backup power generators, but some could only muster a skeletal workforce due to all the homes destroyed and roads left impassable. Sustained island-wide power loss further slowed the resumption of production. Soon hospitals around the globe began to run out of essential supplies like intravenous bags, the preferred method for delivering everything from chemotherapy to painkillers. Because of strict manufacturing specifications, other companies could not close the break in the supply chain quickly. The shortage caused the cost of intravenous bags to skyrocket by 600 percent. With access to bags limited, hospitals turned to syringes to deliver some medications, which, in turn, caused supplies of syringes to dwindle. A year and a half after the storms, drug and medical supply shortages persisted.

Historically, management of supply chains has fallen largely to middle managers and operations teams. That has begun to change, with both corporate and government leaders taking greater interest in protecting supply chain integrity. Just as with climate services, an "arms race" has begun to materialize for information on supply chain disruption.[10] But as one supply chain expert concluded after decades of analysis, "Most companies don't know where their goods are. Full stop."

Understanding how to reinforce or "fatten" supply chains in the face of worsening climate extremes is a vital step for reducing the economic shocks from catastrophic events. In particular, assessments should account for the fact that multiple simultaneous climate impacts may compound the demand for materials. Even a single event can create a "bottleneck" that slows flows through supply chains, when, say, production of an essential component is disrupted or a distribution channel becomes blocked. Research has shown that supply chain disruptions driven by concurrent events may pose greater threats to large wealthy urban areas than a single weather disaster that directly hits a city.[11]

To ensure continued access to the most essential goods, including health supplies, governments may need to step in. During the COVID-19 pandemic, hospitals again ran short of intravenous bags. Because Puerto Rico's still-damaged electricity grid impeded the manufacturing process, the U.S. Food and Drug Administration had to intervene to address the expanding shortages by giving the bag manufacturers priority access to the available power.

The supply chains for medical supplies aren't the only ones that need protecting. Both the pandemic and the impacts of climate change have exposed weaknesses in another vital supply chain—food. As the disease spread across the planet, so did food shortages. When workers fell ill or could not travel, harvesting and food packaging ground to a halt.[12] Rice-exporting countries instituted quotas and curbed or stopped exports. Shortages caused richer countries like Taiwan to dip into their stockpiles, while poorer countries saw prices rise. Reductions in food supplies worsened hunger in Liberia, which depends on imports from India for close to 100 percent of its rice. Hunger threatened to kill millions more than did the pandemic.[13] The UN World Food Programme estimated that the pandemic would cause the number of people facing severe hunger to grow by over 120 million.[14]

It's worth taking a step back to understand how breaks in the food supply chain can turn critical so quickly. Just five countries—Argentina, Brazil, China, India, and the United States—produce about 60 percent of the world's food. Within these countries, food production occurs primarily in concentrated regions. For example, five states in the American Midwest produce 60 percent of U.S. corn, while five states in northern India produce close to 90 percent of that country's wheat. Moreover, much of that food passes through trade "chokepoints," locations like the Panama Canal, U.S., Gulf of Mexico ports, and Brazil's road network.

The geographical concentration of food production makes the global food supply chain increasingly vulnerable in the face of climate extremes. In addition, because world agricultural markets remain finely balanced between supply and demand, running on a just-in-time basis, a local or bottleneck disruption can drive a change in the price of a commodity that reverberates around the globe. When food security is threatened, so is social stability.

Rapidly rising food costs can put sharp pressure on government authorities and, for those unable to respond, can combine with other factors to create explosive conditions. In 2008, for instance, drought in Asia and Australia hit rice growers hard, causing a global shortage of rice. That shortage helped send rice prices to record highs, triggering food riots everywhere from Senegal to Mexico. Likewise, a severe drought in Syria worsened by climate change ravaged agricultural livelihoods and triggered the mass displacement of farming and herding communities. These stresses ultimately added fuel to the protests that erupted in 2011.

Such threats will only grow as the planet warms further. Researchers have found that the risk of simultaneous severe droughts in the world's major breadbaskets could double from 2041 to 2070 even if carbon emissions are reduced significantly.[15] Meanwhile, rising temperatures are speeding up the rate of land degradation and soil loss, and higher concentrations of carbon dioxide in the atmosphere can reduce the nutritional value of crops and harm livestock. All of these climate impacts could reduce yields of essential crops by up to 30 percent by 2050, the Global Commission on Adaptation has warned.[16]

To cope with these challenges, countries need to gain a greater understanding of supply chain risks. They need to identify both where links may break and how they can bolster redundancy. Most importantly, they should try to buffer the strain that a sudden break can cause. For many nations, that will mean turning to the same strategy used by some countries for medical goods and other essential supplies—stockpiling.

Step three: Stockpiling

Stockpiling of essential goods, including food, will need to expand to lessen the severity of shocks caused by multiple disasters. The pandemic revealed the hubris of moving away from stockpiling medical supplies. For example, before the pandemic, Britain and France had assumed that just-in-time contracts and the smooth workings of the open market would ensure access to necessary goods, most likely coming from China. Belgium even destroyed tens of millions of expired masks and never replaced them. And an effort by European officials to build a regional stockpile in 2016 fizzled after France, Britain, and other countries claimed they were prepared. Yet China's export restrictions imposed when the coronavirus began its deadly spread left countries scrambling.

Finland stands out as one country that was ready for the supply disruptions caused by the pandemic. Given decades of worry about threats coming from Russia, Finland's neighbor to the east, "It's in the Finnish people's DNA to be prepared," said the head of Finland's National Emergency Supply Agency.[17] Starting in the 1950s, Finland amassed large amounts of surgical masks and other medical supplies, in addition to oil, grains, agricultural tools, and raw materials. In contrast, its other neighbors, including Norway, Sweden, and Denmark, had largely abandoned the stockpiles they had created during the Cold War era.

Another nation that takes stockpiling seriously is Switzerland, in part because it produces little of its own food. Switzerland maintains one of the largest stockpiles in the world, storing enough food to feed all its citizens for several months. Prompted by a particularly bad storm season in the United States in 2015, Switzerland updated its stockpiling strategy to protect against supply chain disruptions, ramping up caches of food staples, such as sugar and cereals, along with medical supplies in case of a pandemic or bioterrorism incident.

Considering the shortages and disruptions exposed by both the pandemic and the increasing number of climate- and weather-related events, the rest of the world needs to become more like Finland and Switzerland. Some have already started. President Abdel Fattah Al-Sisi of Egypt, the world's biggest wheat importer, ordered expansions in his country's stockpiles during the pandemic. Creating stockpiles, however, will pose major challenges for poorer countries, some of which lack any reserves whatsoever or have only modest ones. For example, in Africa, a region deeply challenged by food insecurity, Côte d'Ivoire, Guinea Bissau, and Senegal had no food stockpiles, while Mali had only around three days' worth of food for its population according to a report in 2016.[18]

Stockpiles of food and essential supplies can ease the jolts and shocks that climate change impacts will bring, lessening the threat from supply chain blockages and therefore strengthening social stability. China, the world's largest importer of food with a population of 1.4 billion, feels the pressure to ensure food security. Many Chinese can still remember the Great Famine that gripped the country from 1958 to 1962. In 1990, the country created a national stockpile for grains. When the coronavirus began its spread, Chinese officials asserted that the country's wheat and rice stores could feed the Chinese population for a year. In 2020, when a series of shocks battered the nation, farmers began withholding products from market as a result of the pandemic, a pest outbreak in the southwest, drought in the north, unprecedented typhoons in the northeastern provinces, and historic flooding in the east that heavily damaged crops in Jiangxi, China, an area known as "the land of fish and rice." Still, China kept its population fed. The Chinese government increased its grain imports by close to 30 percent from a year earlier and dipped into its strategic reserves.

As we will see in Chapter 3, stockpiling on a regional basis can help further reduce some of the overall administrative costs. But it's not just goods that people need to weather climate impacts. They also need a helping hand.

Step four: Planning to surge

2017 was a grim and disaster-filled year in the United States. Three of the top five costliest hurricanes in history, Harvey, Irma, and Maria, made landfall in quick succession. Then large areas in the state of California ignited, leading to the most damaging wildfires in the state's history up to that point. These four disasters alone affected 47 million people, close to 15 percent of the country's population. Because they occurred in rapid sequence, they also placed unprecedented demands on the federal government to respond with emergency aid and help with recovery. Ultimately, the federal government was forced to pony up $300 billion for relief efforts. By May 2018, close to five million American households had registered for emergency assistance, more than the previous ten years combined.

Scientists have determined that climate change heightened the severity of the brutal string of disasters in 2017. This sequence of catastrophes offers a sobering glimpse of what the future will bring, as climate change fuels stronger storms and other weather-related events. Nowhere was this more obvious than in Puerto Rico.

Puerto Rico sits about 1,000 miles southeast of Miami, Florida. For at least the decade before 2017, Puerto Rico's economy had spiraled downward, its government amassing huge debts and defaulting on obligations. The territory's government had skimped on investments in bridges, roads, highways, and other infrastructure like wastewater management. Squeezed for funds, Puerto Rican officials chose to stockpile few emergency supplies, reasoning that strong hurricanes had struck infrequently in the island's history. Authorities never imagined that a set of storms could cause widespread power outages lasting almost a year. But that's exactly what Hurricanes Irma and Maria did within a two-week span.

Maria, a Category 5 storm and the stronger of the two, was the largest hurricane to make landfall in Puerto Rico in almost a century. Maria's fury damaged or destroyed 300,000 buildings. Roads filled with tons of debris and crumpled under floodwaters. Maria even wrecked the headquarters of the Puerto Rico Emergency Management Agency. The storm obliterated the island's entire electricity grid and knocked out 95 percent of Puerto Rico's communications capabilities, triggering the second longest blackout in world history. The longest was the outage in the Philippines after Typhoon Haiyan in 2013. Without power, Puerto Rico plunged into darkness. FEMA immediately embarked on what would become one of the largest recovery efforts, and one of the more troubled, in its 40-year history. As staff deployed from one disaster to the next in 2017, they often arrived exhausted and not in the best condition to serve.

After the U.S. government bungled the response to Hurricane Katrina in 2005, the U.S. Congress directed the creation of the Surge Capacity Force to deploy assistance rapidly in response to catastrophic events. This surge force can call up workers throughout the federal government within 48 hours. Prior to 2017, FEMA had never resorted to using these surge capabilities. But with its own staff stretched across 32 other disaster locations, it used the surge force for the first time to tap workers from across the government to go to Puerto Rico's aid.

Still, FEMA struggled, and relief efforts in Puerto Rico stalled. The U.S. Department of Defense ended up assisting by supporting delivery of medical care as well as mortuary services. Troops installed large generators to get the island much-needed power. The military's humanitarian relief effort grew so large that some military leaders believed the drain on resources may have reduced the readiness of units and commands to conduct future operations.

In examining what went wrong with the response to Hurricanes Maria and Irma, the Government Accountability Office, the federal government's watchdog, found that sequential and overlapping hurricanes, coupled with the California wildfires, had laid bare the challenge of responding to concurrent large disasters. As Caribbean climate scientist Michael Taylor wrote, "Irma and Maria threw out the notion of planning and preparedness based on the expected and the familiar."[19] With experts predicting that storms like Hurricane Maria will go from being one-in-800-year events to one-in-80-year events by the end of the century, while other extremes like flooding, heat waves, and wildfires will intensify and occur more frequently, emergency management needs massive strengthening. Based on my experience, it's clear that the United States needs to improve its preparedness at all levels of government, including its capacity to respond.

This was brought home to me by U.S. Coast Guard officers after they participated in the DHS task force on climate change. Initially skeptical about climate change and its implications for homeland security, they later expressed concern over whether the Coast Guard had the capacity to respond to the concurrent calamities that climate change would bring. They explained that, after the 2010 Haiti earthquake, a very high percentage of their fleet responded to the disaster to provide immediate humanitarian assistance. With temperatures rising in the Arctic, the officers also recognized that commercial traffic would increase in the Arctic Ocean as sea ice rapidly shrank, leading to greater demand for search and rescue, oil spill response, and patrolling of once-inaccessible areas. If multiple disasters struck at once, they concluded, the nation lacked "the cavalry to ride to the rescue."

Part of having a well-equipped cavalry is ensuring adequate training and resources. In 2015, the same year that saw 68 firefighters die while on duty in the United States, I chaired a meeting about wildfire at the White House. Twenty fire chiefs from the western United States gathered to discuss ways to reduce growing wildfire risk. Since 1990, 60 percent of new homes in the United States have been built next to or within wildlands, otherwise known as the wildland-urban interface (WUI). The WUI contains 46 million single family homes and more than 120 million people, and much of it is threatened by wildfire. As we went around the conference table, fire chief after fire chief described how in recent years the nature of wildfires had changed. They said the blazes created their own weather patterns, with embers flying further and fires burning hotter. They lamented that their firefighting crews lacked both the training and the resources to combat these new, bigger conflagrations. Then-Vice President Biden stopped by toward the end of the meeting. In his concluding remarks, he asserted "I can't prove any one fire is a consequence of climate change. But you don't have to be a climatologist, you don't have to be a nuclear engineer to understand that things have changed, they've changed rapidly. The bottom line is your job is getting a hell of a lot more dangerous."[20]

Unfortunately, the U.S. Congress's current funding model for disaster response—opening the federal purse to provide most of the aid *after* disaster strikes—does not adequately address the capacity needs. Rather, it leaves government agencies starved for sufficient funding to prepare. The pandemic provided a graphic example of the cost of chronic underfunding: inadequate investments in state and local public health left authorities scrambling for supplies and capacity to curtail the spread of COVID-19. The reactive funding model also creates a moral hazard for local communities, encouraging them to skimp on preparedness investments knowing that, when disaster hits, the federal government will come to the rescue to bail them out.

Preparedness for a future filled with more extreme events requires strengthening connective tissue so that local communities have the knowledge, resources, and incentives to better prepare themselves. Leaders need to imagine a disaster-laden future, requiring the deployment of multiple teams to help communities simultaneously when, in the United States for example, wildfires ravage huge swaths of California at the same time as floods sweep through the Midwest and multiple hurricanes hit both the Gulf and East Coasts, all of which occurred in the midst of the pandemic in 2020.

With the threat of more devastating storms, heat waves, and other climate change impacts continuing to emerge, and with new pandemics predicted to erupt in the future, individual countries, disaster agencies, and communities will be hard pressed to cope on their own. Some promising approaches for providing extra help and resources already exist. Use of volunteers and surge capabilities can bolster the ranks of emergency responders. For instance, in Australia, emergency responders, both career and volunteer, provide surge capacity by relieving local workers and lending essential expertise.

The city of Paris in 2018 created its own network of citizens to assist emergency personnel during disaster response and recovery as part of its Paris Resilience Strategy. Community-based training by both Paris firefighters and civil organizations equips the city's residents with crisis management skills. This civilian network provides surge capacity to the municipal workforce, and if needed, it can be deployed to assist other public authorities in the event of widespread disasters. Additionally, the city of Kampen in the Netherlands has trained more than 250 volunteers to manually operate flood barriers during emergencies as part of the city's flood defense strategy.

Australia and the United States also have interstate mutual aid arrangements for firefighting. But the success of any mutual aid system rests on the ability to provide assistance at the time it's needed. As catastrophes multiply, these agreements come under increasing strain. Indeed, they already have under the stress of COVID-19. In the United States, the sheer size of the pandemic and the mishandling of supply chains by the federal government pitted states against one another as they competed for resources. Similarly, the unprecedented widespread destruction of the 2017 and 2018 California fire seasons exposed weaknesses in the state's mutual aid system, causing Los Angeles County Supervisor Janice Hahn to conclude it was essentially "broken." She urged greater investment in local firefighting capability.[21]

"Going it alone" in an age of consecutive, concurrent, and compounding disasters, however, will prove expensive and could worsen the risk to communities. Mutual aid agreements can still play an important role if they account for the changing circumstances by planning for heightened demand for services, setting standards for harmonizing training and protocols, and clarifying the limits of any agreement, including whether volunteers enjoy the same legal immunity as career personnel. With more precise agreements, all parties can limit surprises.

Getting ready for multiple disasters

Going forward, planners, including everyone from emergency managers to private sector leaders, will need to tackle the enormous problem of coping with multiple disasters at the same time, something that remains poorly understood and largely ignored. Just as Maria surprised FEMA, so did the pandemic. Eric Letvin served on the National Security Council at the White House with me. He is a civil engineer and lawyer who has held senior positions at FEMA and speaks passionately about the need for predisaster risk reduction. When the pandemic hit in spring 2020, Letvin lamented that FEMA "never imagined having to respond to all 50 states at once." The preparedness task is made even more difficult because steps taken to respond to one disaster may actually multiply the risks from another. Take "cooling centers" where people can shelter from extreme heat events. Designating such centers is a common strategy for responding to heat waves. However, during the pandemic, those shelters could increase the danger of infection, because the coronavirus spreads far more easily indoors than outside.

Or consider the potentially deadly combination of pandemics and floods. As cases of COVID-19 soared, the city of Miami, Florida, converted its convention center into an emergency hospital with 450 beds. But the convention center sits within the mandatory evacuation zone for a Category 1 hurricane. If a hurricane did make landfall, patients too sick to be easily moved would have been at greater risk of injury or death. That's why, when the United States needed additional hospital beds for COVID-19, FEMA urged local officials to locate temporary medical facilities away from areas at risk of flooding.

Fire departments in the American West also faced a difficult choice in the spring of 2020: how could they protect firefighters during a pandemic? Fire officials worried that a large fire would expose both evacuating citizens and firefighters to greater risk of contracting COVID-19. To reduce the threat, they decided their firefighting strategy should aim to stop fires early, before they could spread. Taking this approach would cut back the number of people evacuating and the need for base camps for firefighters to use.

This decision to act aggressively and early to contain fires added even more fuel to the next fire, however. It let vegetation that would otherwise burn accumulate, which in turn could lead to a bigger blaze in the future. Firefighters know that letting wildfires burn improves forest health in the long term, but, in the face of COVID-19, "That is not even an option," according to Mike Morgan, director of the Colorado Division of Fire Prevention and Control. "The risks are just too great," he remarked.[22] Despite this planning, record-setting wildfires exploded out of control in California under climate-charged heat extremes while the pandemic raged.

As these examples show, simultaneous disasters complicate essential emergency responses like evacuations of people and mass sheltering. In the face of multiple disasters, those planning to manage risk need to imagine the additional resources and capabilities they will require to effectively respond. "Stress testing" disaster plans and strategies by seeing if they hold up to imagined scenarios will unmask areas demanding reinforcement. Similarly, hosting local, national, and international emergency response exercises that include leaders from all levels of government as well as the private sector and NGOs could improve future planning, identify resource gaps, and build relationships among responders. As climate change impacts cause more shocks, communities and governments should imagine and plan for concurrent, consecutive, and compounding disasters.

Harvest the "no more" lessons from every disaster

Every disaster offers a "no more" lesson that communities can use to do better. Only the really big events—the "no more" moments—however, typically prompt government-led investigations to determine how things went so terribly wrong. For other, more routine disasters, policymakers must dig through a hodge-podge of media accounts, audits, studies, and "after-action" reports to assess what went awry and how policy could drive better results in the future.

Adding to the challenge is that disaster response agencies rely on self-analysis to examine their performance. But self-evaluation can lead to watered-down results as report drafters seek to avoid finger-pointing or upsetting their bosses. Moreover, some investigative reports are held under wraps or released only after the disaster has faded from memory. Not surprisingly, many disaster reports end up gathering dust. As a result, the same mistakes keep happening.

Establishing a system that ensures ongoing improvement could lead to better results. The United States' experience with the National Safety Transportation Board (NSTB) proves that.

Created in 1967, the NSTB is a small, independent U.S. government agency. Its purpose is to determine the likely cause of transportation accidents and then propose corrective steps that could prevent such calamities from recurring. It has no authority to require implementation of its recommendations, but it can advocate for them, something it has done very effectively, even going so far as publishing an annual list of its "Most Wanted Safety Improvements" to capture public attention. In other words, the NSTB turns every accident into a "no more" moment, believing that "out of tragedy, good must come," in the words of the board chairman.[23] As of 2018, the NSTB had issued close to 15,000 recommendations, of which 82 percent had been successfully closed. Many credit the NSTB with making commercial aviation as safe as it is today.

Creating a similar board to investigate smaller disasters, as the U.S. Congress proposed on a bipartisan basis in 2020, could add an important tool to the preparedness tool chest. Combing through every disaster in search of ways to improve will allow responders and the communities they aid to learn from past mistakes and achieve better outcomes in the future. Regular, rigorous, and independent review of disasters could improve preparedness efforts.

A failed response opens the window for trouble

When governments fail to respond quickly or effectively to multiple disasters, the risk of food and water shortages, disease, and loss of livelihoods rises. In addition, if the public doubts the commitment of authorities to help survivors, that loss of trust can undermine support for the government. That's when opportunists, such as armed insurgents, organized crime, and terrorists, can move in to capitalize on the citizenry's misery. As the pandemic revealed, failed, weak, or sluggish responses in the immediate aftermath of disaster provide inroads for bad actors to use humanitarian aid to recruit members and expand territory.

While the Nigerian government struggled with COVID-19, for example, the Islamic State in West Africa took advantage by ambushing government forces. In Mali, al-Qaeda-affiliated jihadists murdered soldiers. On the same day, Islamist extremists seized a strategic port in Mozambique, hoisting their flag as proof of their control, and Boko Haram ambushed Chadian soldiers. In South America, Colombian armed groups used the crisis to murder land rights activists. Other groups used the pandemic to recruit, spreading propaganda and conspiracy theories. Houthi rebels in Yemen told young prospects that it is "better to die a martyr in heroic battles than dying at home from the coronavirus."[24] In Somalia, al-Qaeda-affiliated al-Shabab fighters blamed the pandemic on "crusader forces who have invaded the country."[25]

Similarly, climate change–related disasters can destabilize governments and make people more vulnerable to bad actors. Experts say that the West Pakistani government's poor response to the 1970 Bhola cyclone in East Pakistan (now Bangladesh) was "the principal reason" for the civil war that led to the declaration of Bangladeshi independence in 1971.[26] Political leaders in East Pakistan accused the West Pakistani government of "gross neglect, callous inattention, and utter indifference."[27] The humanitarian crisis added fuel to the independence movement, and within months, civil war had erupted.

Climate impacts disrupt critical human and natural systems, triggering crop failures, blackouts, and destruction of infrastructure. As these effects cluster and compound, breakdowns in security, governance, and economic productivity become more likely. In larger developing countries, the loss of a few percentage points of gross domestic product (GDP) every five to ten years can retard development over the long haul. Even advanced industrialized nations will find climate impacts a disruptive drag on their economies. Bolstering disaster response globally is thus critical for preventing these breakdowns and helping communities and people to cope with the worsening impacts of climate change. The real solution to improvement is to move from simply responding after disasters strike to managing and reducing risk beforehand by boosting preparation. Communities and nations need to be proactive to assemble tools to respond effectively, including reinforced supply chains, stockpiles, and surge capacity. Managing risk means understanding that the past no longer guides the future and that consecutive, concurrent, and compounding disasters require deep investment in emergency preparedness to avoid severe disruptions. Governments that fail to prepare may face "the risk of an alliance cemented by despair," as their populations seek help wherever they can find it.[28]

A tragic story in India during the pandemic offers a cautionary reminder of the challenges that lie ahead for all of humanity. As COVID-19 spread in May 2020, Cyclone Amphan hit. It was the first time India confronted two simultaneous disasters.[29] In advance of the storm's landfall, some Indians, fearing infection from the coronavirus in crowded shelters, refused to evacuate. Some who did not evacuate were crushed as their homes collapsed or were killed by falling trees and debris. The cyclone's high winds and torrential rains also ruined newly sown crops that were needed to feed communities through the next season. Roads washed away. Hospitals, heavily damaged from flooding, could no longer treat coronavirus patients. State Premier of West Bengal Mamata Banerjee said that she had never witnessed such a bad disaster.[30] The storm damaged or destroyed some 2.7 million homes. As one local Indian official reported, "The coronavirus has already taken a toll on people. Now the cyclone has made them paupers."[31]

CHAPTER 3

PLAN ACROSS BORDERS

Cauvery, the river that never fails, even if the sky does.

—*Kadiyalur Uruttiranganannar, Tamil poet living in the 3rd or*
4th century AD

W hen I heard the news in early 2020 about a highly contagious disease emerging from China, I remembered a discussion from years earlier. The Department of Homeland Security's (DHS) task force on biological threats had debated whether DHS, as the agency responsible for U.S. borders, could stop an infectious disease before it entered the country. It was the first time I had heard the expression "borderless disaster."

But then President Trump did something that the task force had never discussed. On March 12, 2020, he announced that the border with Europe would close in 48 hours. Americans quickly filled every flight headed home, paying whatever it took to beat the deadline. Once those planes landed, passengers stood shoulder to shoulder for hours as DHS processed their re-entry into the United States. The massive, dense crowds of returning Americans likely accelerated the spread of COVID-19 within the country's borders. On March 11, 2020, New York had detected approximately 220 cases of novel coronavirus. Two weeks later that number had jumped to 25,000, with genetic tracing showing that most of the infections had come through Europe, rather than directly from China, into the United States.[1] By jamming thousands of people at the border checkpoints simultaneously, Trump's border closing supercharged the spread of the coronavirus from Europe to the United States.

The essential point the Trump administration missed is that by then, there was nothing it could have done that would have prevented the disease from reaching the U.S. shores. In fact, examination of blood donations shows that the virus had arrived in the United States as early as December 2019. From its start in China, COVID-19 has crossed every international border and many internal borders throughout the world, despite efforts to impose travel bans and other restrictions. As we saw in Chapter 1, what did work to stop the disease's spread inside nations once it had entered was strong leadership driving early action to detect and contain the virus.

Similarly, the borderless disasters caused by the accumulation of greenhouse gas emissions wash across all borders. Like pandemics, climate disasters do not honor the jurisdictional and geopolitical boundaries that humans have erected over time to organize themselves. The same type of climate change–fueled droughts and extreme heat that touched off devastating wildfires in Australia are likewise driving unprecedented blazes everywhere from California to Siberia. Rising seas will swamp Florida just as inevitably as Bangladesh and the Marshall Islands. As countries and communities try to address climate extremes within their borders, their actions can affect those just across the line.

Take efforts to address flooding. When one city erects a seawall, it will likely push water onto adjacent shores unless communities collaborate. Likewise, a town's failure to address climate-worsened flooding can spill over to neighboring communities. For instance, one study that modeled the responses of three neighboring coastal counties in the San Francisco Bay Area to flooding caused by sea-level rise found that unilateral action could disrupt activities in the other two counties. If one county chose not to protect its shore-line and instead allowed its roads to flood, that would result in cascading traffic delays for the other two counties, increasing the time commuters spent on the road by over 10 percent. On the flip side, if one county took action to protect its coastline, the other communities suffered more flooding that delayed traffic as well.[2] The same holds true for areas where rivers are prone to flooding. As the state of South Carolina's Floodwater Commission succinctly stated, "It does little good for one local jurisdiction to have high-quality plans, if the upstream jurisdiction does not."[3]

Unilateral actions to reduce the risks of climate impacts can even carry unintended global consequences. Consider the construction of desalination plants to combat drought-caused water scarcity or the reliance on more air conditioning to deal with greater heat extremes. As we will see in Chapter 6, both of these measures can contribute to rising temperatures by increasing emissions when powered by fossil fuels. This, in turn, exacerbates climate extremes globally. Efforts such as construction of a dam to secure water for clean hydropower by one country can likewise reduce freshwater flows to countries downstream.

What makes borderless disasters so difficult to combat, however, is that the same borders that fail to stop pandemics and climate impacts are also major impediments to the collective action imperative to reducing the threats. The division of the world into sovereign states with national and subnational borders, drawn in many cases during colonial times, presents "intractable barriers to cooperation and collaboration."[4] Without joint action, a nation that fails to contain a pandemic can seed new outbreaks in other countries that have been more successful and responsible. A nation that fails to cut back its CO_2 emissions can cause global temperatures to keep climbing and bring worsening climate extremes to all.

Tackling problems like pandemics or climate change within the framework of traditional jurisdictional boundaries means that policymakers continue to treat these challenges like matters of domestic or local concern, rather than the transboundary threats that they are. Breaking down these barriers requires deep focus on cross-border solutions. For example, the climate change problem of "too little and too much water" demands transboundary consideration of evolving conditions in river basins and ocean fisheries. Crafting risk reduction efforts that stretch across regions—such as stockpiles, insurance risk pools, and climate forecasting and early warning systems—also strengthens disaster preparedness. Likewise, greater cross-border cooperation—not less—is required to solve the most urgent transborder challenge of all, climate-induced migration.

Too little and too much water

Too little and too much water is one of the major recurring themes of climate change. Drought is the obvious example of too little water. Sea-level rise and extreme rainfall are examples of too much water. Sometimes both problems can occur in the same place, like when torrential rainfall hits soil, hardened to concrete by drought, causing more serious flooding. Dealing with water challenges gets more complicated when the impacts cross jurisdictional boundaries, which happens much of the time.

Many nations share a single water source. As climate change intensifies and the effects on water become more pronounced, jurisdictions will need to collaborate to find solutions. From the Rio Grande that divides the United States and Mexico to the mighty Nile River that flows through 11 countries in East Africa, the globe's experience with sharing river basins both illustrates the problems and offers solutions.

Nations and communities have shared river basins ever since humans formalized geographical borders. There are some 310 international river basins that form or cross national borders. These basins supply close to 60 percent of the total freshwater resources on the planet. Historically, nations have usually worked out their differences without having to resort to war. Countries sharing rivers have developed approximately 300 agreements to govern water rights. These accords have proven vital to ensuring peace. Still, large gaps remain in coverage. More than half of the basins lack an international agreement. Few of the multilateral treaties include all of the nations within the basin or cover the entire geography of the basin. Only a third of the areas with shared river basins have a formal governing body to conduct joint water management.

In 2014, the UN Watercourses Convention came into effect when a little over 30 nations, mostly in Africa and Europe, ratified it. The convention establishes the first global framework for cooperation over water resources between countries. But without wider adoption, including in the Americas and Asia, its provisions do not govern some of the planet's biggest water disputes.

In recent years, the world's largest river basins, including the Nile, the Mekong, and the Indus, have come under rising pressure as individual nations seek to siphon off larger shares of water. Some of the demand for water comes from agriculture, which consumes nearly 70 percent of the world's fresh water. Some of the demand comes from the desire for hydropower to generate electricity for expanding populations. According to Aaron Wolf, an international water treaty expert, "It's a blueprint for disaster when a country upstream wants to build a dam and has no agreement with the country or countries downstream."[5] Climate change adds to these water woes.

The impacts of climate change, such as reduced rainfall, earlier snowpack melt, and increased flooding, can have major effects on water flows. Drought and higher temperatures can cause water levels to decline. Rising temperatures can degrade water quality through algae blooms. Melting glaciers can initially cause flooding but eventually lead to water scarcity. Yet when nations negotiated most of the world's transboundary water agreements, the threat of climate change received no consideration. Thus, the agreements lack adequate mechanisms to adjust to changing climatic conditions. The inability to easily modify these pacts in the face of accelerating climate extremes could drive future conflicts, especially if nations insist on fixed contractual arrangements based on past conditions. Going forward, therefore, these accords will require renegotiation to address changing conditions and new scientific knowledge about future scenarios.

In a few cases, countries have successfully updated their transboundary water accords, at least temporarily. In 2017, for example, the United States and Mexico adjusted their transboundary treaty that was originally signed in 1944 for the Colorado River basin. That basin, like so many others, faces expanding water demands and declining water flows. The two countries agreed that when water shortages occurred, possibly as a result of climate change, both countries would make voluntary cutbacks on water demand for the next nine years. In addition, other nations that share rivers, including the Danube and the Rhine, have developed adaptation strategies or are in the process of doing so.

Despite this progress, the governments controlling many river basins have not begun the vital task of accounting for climate change, and the consequences of that failure are now apparent. One treaty that has already come under stress is the 1960 Indus Water Treaty (IWT).

Scholars frequently cite the IWT as an example of the importance of reaching transboundary water agreements. Brokered by the World Bank, the treaty divides the six major rivers of the Indus basin between India and Pakistan, two bitter adversaries. Pakistan gained rights to most of the water in the region's western rivers, which run through Indian-administered Kashmir, an area over which Pakistan asserts rights as well. Pakistan depends on the Indus as its primary source of fresh water, with the basin's rivers supporting 90 percent of the country's agriculture. In return, India gained unlimited rights over the eastern rivers in addition to some rights over the western rivers primarily for power generation and navigation, as well as limited irrigation.

The treaty has long provided an area of cooperation between India and Pakistan. But the region is now particularly at risk from climate change, and there are increasing concerns the treaty may not hold as conditions change. Between 1962 and 2014, the Indus basin experienced a 5 percent decrease in flows. The International Monetary Fund listed Pakistan third among countries confronting severe water shortages in 2018. Climate projections predict that river flows could drop dramatically in the future as the glaciers in the Himalayas, which feed the Indus waters, rapidly melt and disappear. With the IWT containing no mechanism to deal with climate change impacts, it may prove weaker than many had hoped.

In fact, in 2016, India startled the world by announcing it might pull out of the treaty, in part over Pakistani-based terrorist attacks on an Indian military post. Then in 2019, under pressure from farmers demanding more water for agricultural use, the Indian prime minister vowed to end the treaty unilaterally. Pakistan responded by calling the proposed unilateral termination an act of war. For Pakistan, loss of the Indus waters would drastically curtail the nation's access to fresh water and its ability to irrigate cropland to produce food, forcing it to rely on food imports to feed its rapidly growing population. Mention of war carries weight. The countries, both nuclear powers, have already fought four wars since India gained its independence in 1947.

Climate change threatens to upend more than just international water agreements like the IWT. It also stresses agreements reached within countries. Take the Cauvery River, which runs about 800 kilometers (500 miles), making it the fourth largest in India and a popular subject for poets. From ancient times to the present, the river has supplied water to those living along its banks. The state of Karnataka and its downstream neighbor, the state of Tamil Nadu, the most urbanized state in India, have argued over the Cauvery River's water for more than 120 years. Both states lay claim to the water, and both need it for their expanding populations.

The city of Bangalore in the state of Karnataka is considered India's Silicon Valley. In 1950, Bangalore had a population of

150,000. In 2018, over 11 million called it home, and by 2031, the city's population is expected to grow to 20 million. Bangalore relies on the Cauvery River for drinking water, which it pumps uphill for 100 kilometers (62.1 miles) at the monthly cost of about $6 million in electricity. The Indian state of Tamil Nadu depends on the Cauvery for agriculture, including water-intensive rice cultivation, and industry. Bigger populations and changing rainfall patterns that lower river levels have caused significant water stress in recent decades between the two states. In periods of adequate rainfall, tensions subside, but when rain runs short, tensions flare.

In the 1990s, the Indian government created a tribunal to resolve the issues. Over the years, protests about allocations have included hunger strikes by local officials and suicides by farmers. In 2016, after drought severely reduced water availability in both states, the Indian Supreme Court ordered Karnataka to release water to Tamil Nadu. Protests erupted in Bangalore. Authorities brought in 15,000 law enforcement officers to restore order. An actor named Simbu started a movement to quell the heightened tensions by urging people from Karnataka to offer a glass of water to their friends in Tamil Nadu to prove that they could share the water peacefully.

In 2018, the Supreme Court ordered the creation of the Cauvery River Management Board to direct the sharing of the river's waters. The court's final judgment in 2019, issued after years of appeals and prior orders, declared that no state could deviate from the division of waters ordered by the court and that this new sharing scheme would apply for 15 years. Researchers say, however, that the dispute remains "far from resolved."[6]

Past attempts to settle the controversy based on the historical or current river flows have not worked. Climate change will make future negotiations even more difficult by continuing to alter water levels in the Cauvery River. Any lasting decision will need to account for the risk of "too little" water caused by climate change. Contrary to what the poet Kadiyalur Uruttirangannanar wrote over 17 centuries ago, the Cauvery can fail when the rains stop (Figure 3.1).

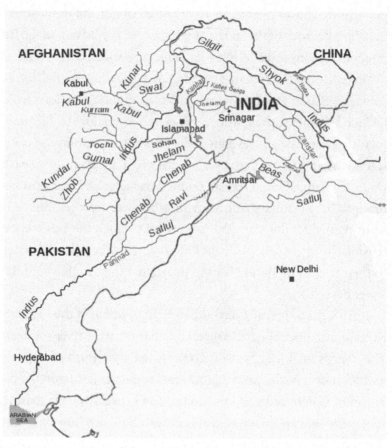

FIGURE 3.1. Map of the transboundary Indus River Basin, which is governed by the Indus Water Treaty between Pakistan and India.
Source: Wikimedia, 'Indus River Basin.'

To manage river basins going forward, national and local governments will need to reach water-sharing agreements that can adjust to fluctuating conditions, including climate change, increasing urbanization, and population growth. Agreements can no longer safely rest on the assumption that historical flows will resemble future water flows. Similarly, adjacent communities will need to plan together for sea-level rise. One jurisdiction's actions could harm another jurisdiction when, for example, development removes wetlands that protect against storm surge. In the face of the burgeoning set of cross-border risks, more transboundary planning should occur—something both the Organisation for Economic Cooperation and Development (OECD), a conglomeration of wealthier countries, and the European Union have called for.

Where have all the fish gone?

In 1980, I spent a year on Japan's northern-most main island, Hokkaido, teaching English in a small private school in a small coastal city. Hokkaido is one of the least populated places in crowded Japan—and the coldest. While I lived there, snow covered the ground for more than half the year. Surrounded by the Sea of Japan, the Pacific Ocean, and the Sea of Okhotsk, which is wedged between Russia and Japan, the island has accounted for 20 percent of Japan's total fisheries output. Hokkaido is famous for its salmon, and salmon fishing dominates its local culture. The indigenous people of the region, the Ainu, have some 40 words for salmon. Seemingly everywhere I looked, I would see a statue or carving of a bear with a salmon gripped in its teeth. In every noodle shop and tea house, salmon dominated the menu. Celebrations required copious amounts of salmon roe. On New Year's Day, friends delivered thick packets of freshly caught wild salmon as gifts.

But conditions have changed since I lived on Hokkaido. Global warming has shrunk the ice cover on the Sea of Okhotsk by 30 percent.[7] As one of the fastest-warming areas in the world, the sea has experienced profound changes in its salmon population. Salmon are highly sensitive to heat, and when they encounter warmer waters, it throws them off their migratory path. Since 2003, northern Japan's salmon catch has plummeted by 70 percent. With the waters off the coast of Hokkaido warming, salmon have departed in search of cooler waters. Meanwhile, Russia's fisheries have gained. That country has experienced a fourfold increase in its salmon catch. But Hokkaido is not alone in its changing fisheries. Global heating is altering fish migration patterns around the globe.

Oceans cover more than 70 percent of the planet's surface. The fish they contain are the last major wild food source and provide the primary source of protein for over one billion people. Over 95 percent of the people who fish for a living reside in developing countries, with fishing as their major source of income. Many historically important fisheries are now threatened by rising temperatures, declining oxygen content, shifts in currents, and acidification, which have caused fish to migrate, often toward cooler water in the polar regions. Oceans have warmed almost 1 degree Celsius (2 degrees Fahrenheit) since 1850. Researchers estimate the average pace of migration at 37 miles (59 kilometers) a decade.[8] Young great white sharks have moved by an average of 373 miles (600 kilometers) between 2014 and 2020 northward off the coast of California toward cooler waters.[9] Because this redistribution of species has profound effects on those dependent on fish for food and/or livelihoods, the governance of fisheries needs adjustment. Countries in the tropics, like Indonesia and other nations near the equator, have the most to lose since their stocks will only decrease with no new species to replace those leaving.

Fisheries management, like water basin management, still largely assumes that the geographical range of fish stock remains constant. The 1982 UN Convention on the Law of the Sea gives individual countries rights to exclusive use of resources within the economic zones (EEZs) off their coastlines. It also requires nations to ensure that fisheries within their EEZs are not endangered by exploitation. A 1995 compact, the UN Fish Stock Agreement, requires nations to cooperate to protect migratory fish and fish that "straddle" boundaries against exploitation. But neither the UN Convention on the Law of the Sea nor the Fish Stock Agreement addresses the movement of fish stocks beyond territories they have historically occupied or have left altogether due to climate change.

Countries have likewise entered into regional and bilateral fisheries agreements. Like the international pacts, however, fisheries agreements do not yet typically account for fish stock movement in response to climate change. A study of 127 fisheries accords in the South Pacific, Southeast Asia, South Asia, Northwest Africa, Central America, and the Caribbean found that none accounted for the fact that climate change now drives fish stocks out of their historical ranges.[10] Nor did the accords address the risk of overfishing as nations gain and lose fish stocks.

One-third of all fish stocks are already overfished.[11] The risk of overfishing grows as fish stocks begin to cross human-drawn borders in search of cooler waters. Local fishermen and -women may try to capture more of the fish that remain within their nation's EEZs before the fish migrate to other areas. And when the fish arrive in new territorial waters, fishermen and -women in those areas may overfish in the absence of guidance as to sustainable fishing practices for the new species. Thus, newly migrant fish stocks may face "double jeopardy" for overfishing in both jurisdictions. According to one estimate, 23 percent more fish will be crossing international borders by 2060.[12]

Overfishing will threaten fish stocks. This would lead to the loss of a major food source for billions of people as well as the loss of livelihoods for the millions employed in the seafood industry. The encouraging news is that at least some small progress has been made toward agreements that reflect the new climate realities. In 2015, for example, ten countries agreed to prevent unregulated commercial fishing for 16 years in the high seas of the Arctic Ocean, a 1.1-million-square-mile area known as the "donut hole." The area has historically been frozen over, but as sea ice melts in the summer, commercial fishing becomes more attractive. Under the agreement, the nations decided to hold off on fishing until scientific research can inform a management plan.

Consider too the progress that two of the world's great fishing nations, Peru and Chile, have made in resolving climate-stoked disagreements over catches in the South Pacific's Humboldt Current. Drawing cold water from the depths of the ocean to the surface, the current sweeps northward from mid-Chile to Ecuador. With this upwelling comes an entire food chain, ranging from microscopic plankton to sharks, creating a fishery that produces as much as one-fifth of the world's wild fish catch. But climate change can alter upwelling, and for the two adjoining countries of Peru and Chile, it will affect the fish stocks in their respective EEZs.

In 2014, Peru and Chile turned to the International Court of Justice in The Hague to resolve a maritime border dispute regarding their respective EEZs. More recently, with the help of the nongovernmental organization (NGO) the Environmental Defense Fund, Peruvian and Chilean scientists have collaborated to create the Humboldt Current's first-ever early warning system for climate impacts on fisheries in collaboration with Ecuador.[13] The goal is to create a common understanding of how climate change alters their fisheries. By working across borders to obtain more accurate projections of fish stocks, all three countries can better prepare for the uncertainties that climate change brings.

But this is just a start. Fisheries face additional challenges beyond the movement of fish across international borders as a result of climate change. Climate-driven acidification of ocean waters affects spawning grounds in coastal waters and degrades the health of reefs, which, in turn, can reduce the number of fish. Many other agreements that guide the sharing of fisheries will need adjustment in the face of rapidly changing conditions, just as we saw holds true for river basins. The experience of the past is no longer a safe guide for the future.

Banding together

In times of stress, nations and communities can turn inward. They hoard goods and close borders. They reinforce jurisdictional boundaries rather than transcend them. But these inward maneuvers can be counterproductive, because at times working across borders can provide access to more resources at lower costs. That's why, in regions around the globe, some countries have begun to band together to prepare for crises. By jointly stockpiling foods, promising mutual assistance during humanitarian crises, and developing regional early warning systems, these countries expect that they will save more lives and money together compared to going it alone.

The 2020 pandemic's impact on world food supplies shows the benefits of cooperation. The lockdown measures that countries imposed restricted the ability of growers, processors, and marketers to supply food. In some countries, the disruptions threatened to raise food prices, especially in the countries that import most of their food. As we saw in the last chapter, nations have created stockpiles to protect themselves against global shocks. Regional stockpiling of food, however, provides an additional layer of protection, while avoiding the delays that can slow delivery of international humanitarian aid.

The best-known regional food stockpile is the ASEAN Plus Three Emergency Rice Reserve (APTERR). Agreed to in 2011 after years of negotiations by the Association of Southeast Asian Nations (ASEAN) member states with China, Japan, and South Korea, the reserve aims to ensure the availability of rice during food shortages, including ones caused by climate change. The program additionally monitors the availability of essential commodities, so that it can quickly trigger disbursement of food supplies. For instance, APTERR sent an emergency supply of rice to the Philippines in 2020 after a typhoon and an earthquake struck the country, and during the pandemic, APTERR officials placed the stockpile at the ready to alleviate shortages. As COVID-19 spread in early March 2020, Japan contributed 300 metric tons of rice through APTERR to supplement food supplies in Myanmar's conflict-ridden Rakhine State.

The regional stockpile in western Africa got its start after the 2008 global food crisis. The Regional Food Security Reserve of the Economic Community of West African States (ECOWAS) seeks to improve cooperation among countries regarding their national stockpiles, embracing the concept of "three lines of defense"—draw first on local supplies, then national stockpiles, and last on regional reserves.[14] South Asian countries have created their own regional stockpile as well, the South Asian Association for Regional Cooperation (SAARC). With the ability to reach the hungry quickly, these food reserves help reduce the rate of hunger-related illness and the need for costly nutritional recovery programs.

International mutual aid agreements can likewise improve outcomes. The European Commission created the EU Civil Protection Mechanism in 2001 to pool emergency resources for EU countries. The mechanism coordinated the single largest staff deployment to help Sweden battle the biggest fire in its history in 2019. To offset the increasing intensity and frequency of wildfires, the European Commission created "rescEU," a fleet of firefighting helicopters and planes standing at the ready for mobilization.

Global mutual aid pacts for wildfire have worked particularly well in the past because "fire seasons" historically occurred at different times in the Northern and Southern Hemispheres. These arrangements developed organically over time, but in 2017, Australia and the United States formalized an agreement to help each other with fire suppression efforts. In 2018, Australian and New Zealand firefighters traveled to the American West to battle blazes. Returning the favor, the United States, as well as Canada, deployed firefighters and firefighting equipment to help Australia combat the Black Summer fires of 2019–2020. As climate change lengthens fire seasons, however, these agreements need re-examination to avoid cracking under the pressure of events occurring simultaneously in both hemispheres.

Regional climate services and early warning systems could also save money and improve cross-border planning. Yet, according to the secretary general of the World Meteorological Organization (WMO), close to half of the world's countries lack effective climate services and early warning systems.[15] To begin to close the gap, the WMO, with funding from international development agencies, has created 20 Regional Climate Outlook Forums that bring together national, regional, and international climate experts to produce climate predictions in real time to aid decision-making on the ground. Policymakers can use the forecasted climate conditions to inform response plans for different sectors and industries such as public health, tourism, agriculture, and security. Forming the forums recognizes that climate change affects entire regions, necessitating cross-border information sharing and planning.

Regional early warning systems not only provide cost savings to individual governments but also improve preparedness efforts across shared borders. The Regional Integrated Multi-Hazard Early Warning System for Africa and Asia (RIMES) has over a dozen member nations. In Asia, the UN Economic and Social Commission for Asia and the Pacific has created a regional early warning system for drought as well as two warning efforts focused on tropical cyclones. Expanding regional early warning systems to transboundary river basin flooding could save lives and foster greater cooperation among the affected nations.

The success of regional climate services depends, however, on the capacity of the nations and their citizens to receive the information and act upon it effectively. And for many countries that will remain a struggle given that over 50 percent of the world's population, more than four billion people, lack access to the internet. Similarly, the absence of reliable electricity can hinder dissemination of vital warnings and forecasts. Notably, in sub-Saharan Africa nearly 600 million people, making up almost two-thirds of the region's population, lack regular power; this shortage afflicts almost a quarter of those living in South Asia as well.

Since the early 2000s, nations have banded together on a regional basis to obtain better insurance protection against catastrophic risk. Many developing nations lack insurance for disaster losses, which leaves them scrambling for funds after disasters. They can't depend on international donor funding because it may or may not flow after calamity strikes. Without emergency funds on hand, disasters force reallocation of money from other needs, for example, education and infrastructure, to pay for crisis response. All this delays provision of emergency aid, which, in turn, leads to more lives and livelihoods lost. One solution is regional insurance pools, which can get the money out faster.

Sixteen Caribbean nations joined together in 2007 to form the world's first and most well-known multicountry risk pool, the Caribbean Catastrophe Risk Insurance Facility (CCRIF). Hurricane Ivan had swept through the region three years earlier, causing billions of dollars in damage. In Ivan's wake, the affected countries worked with the World Bank and others to create an insurance scheme. Nations in the Pacific and Africa have followed the Caribbean example. In 2007, the Pacific Catastrophe Risk Assessment and Financing Initiative (PCRAFI) launched with 15 countries. In 2012, African nations formed the African Risk Capacity (ARC) to address drought risk.

Because insurance pools diversify the risk across geographic regions, they can offer cheaper insurance and cut administrative costs through economies of scale. To speed up payouts, the pools often use parametric triggers, operating in a manner like forecast-based action discussed in Chapter 2. With parametric insurance, a defined event, say, the level of flooding according to a set of flood gauges or deviation in rainfall from a historical pattern, determines payouts. If the trigger is met, the money is paid. Since the insurance benefit is received no matter the level of loss, the process of verifying and assessing damages is no longer required.

Payouts can occur quickly. Just seven days after Cyclone Gita struck Tonga in 2018, the PCRAFI pool transferred funds to the Tongan government to support disaster relief efforts. By providing rapid cash, the pools allow governments to deliver fast emergency relief to their citizens, diminishing loss of life, malnutrition in children, and deterioration of assets. During a drought in the Sahel in 2014 and 2015, ARC made payouts to three countries, Niger, Senegal, and Mauritania, in January 2015, a month before an international appeal for aid was even launched. The Boston Consulting Group estimates that every $1 spent on early intervention through ARC saves $4.50 that would have been spent in traditional emergency response.[16] Risk pools could likewise play an important role in driving predisaster risk reduction by, for example, offering discounts on insurance premiums for risk mitigation measures.

But, for all their promise, these insurance schemes contain potential pitfalls. If the triggering event is set too high, countries hit with a disaster may still not receive a payout. Some researchers have likened the approach to "acting more like an expensive lottery ticket than a cheap way of purchasing protection."[17] In 2015, the Solomon Islands left the Pacific pool after suffering two disasters without a payout. Similarly, the Bahamas, a member of the CCRIF since its founding, briefly decided to opt out of the pool after having spent $9 million on premiums over the course of a decade and never receiving a payout. But by 2019, the deputy prime minister reached a different conclusion, declaring that the CCRIF is "worth it" after it paid out $12 million within days of Hurricane Dorian striking the nation.[18]

Mustering the money to pay the premiums can pose a barrier. Though international donors have typically carried much of the expense, the costs to national governments can become burdensome enough to cause some countries to drop out. For instance, Kenya left the Africa pool in 2016, citing political pressure to justify the premium. Moreover, the money spent on insurance premiums does not reduce the risk a nation faces. The insurance is a way to transfer risk to another entity, not to reduce the risk itself. To protect themselves against mounting risks from climate change, nations should couple insurance with risk reduction measures. A recent lesson from the coronavirus speaks to the importance of issuing funding focused on preparedness rather than damage.

Setting triggers that do not correspond to the nature of the problem that needs solving can undermine the effectiveness of parametric financing schemes. The World Bank encountered heated criticism when it did just that in launching pandemic bonds to great fanfare in 2017. After the Ebola pandemic killed over 11,000 people in New Guinea and Liberia, the bank wanted to find new ways to funnel cash through the financial markets to developing countries threatened by future pandemics. Like parametric insurance, the pandemic bonds had a trigger—a certain number of deaths. Critics lambasted the bond's design, noting that providing funds after an infectious disease had taken root sufficiently in a country to cause multiple deaths would do little to help countries shore up their failing health systems to stop the spread. Some called the scheme "obscene."[19] Larry Summers, a former World Bank chief economist, termed it "an embarrassing mistake."[20] When the COVID-19 pandemic hit, it was not until late April that the deaths had reached the requisite number to release the funds to 64 of the world's poorest countries.

Banding together to share resources and build greater financial protection through well-designed programs can help communities and nations build resilience to accelerating climate impacts. Cross-border initiatives and planning can reduce the number of people permanently dislodged from their homes as a result of climate change. None of these efforts, however, can prevent climate change–driven displacement altogether. Nations need to come up with a different solution for that challenge.

Survival migrants

In 2009, the year I started at DHS, U.S. immigration officials apprehended about 20,000 children at the U.S. southern border. Over 80 percent of them came from Mexico. Two years later, kids from Central America's Northern Triangle, the countries of Guatemala, Honduras, and El Salvador, started showing up at the border in larger numbers than ever before, eventually surpassing the number of Mexican children. In 2011, there were 16,000 from the Northern Triangle; in 2012, there were close to 25,000; and in 2013, there were 38,000. By 2014, the number swelled to almost 70,000, straining the existing system beyond its capacity.

In 2012, DHS Secretary Napolitano asked me to coordinate DHS's efforts to manage the growing crisis. I visited DHS immigration facilities in Brownsville, Texas, and saw kids sleeping in cells with only space blankets covering them. I toured attractive dormitories run by the Department of Health and Human Services (HHS) that offered an array of social services. Once I had assembled a DHS task force to oversee the process, we worked on getting the kids out of DHS custody and into the hands of HHS as quickly as possible. We also created a public messaging campaign in Spanish advising of the dangers of traveling north and urging families to keep their children at home. We worked with the Mexican consulate to find ways to better address the flow of kids crossing the border into the United States. Still, the river of humanity kept coming.

By 2019, the number of kids from Central America dwarfed the number from Mexico. Eighty-five percent of the children stopped at the border came from the Northern Triangle. For many, their journey involved riding on top of "La Bestia" or "The Beast," the train that traveled north through Mexico, leaving them vulnerable to violence, theft, and human trafficking (Figure 3.2).[21]

One driver of the large flows of migrants from Central America is climate change. The Northern Triangle countries are among the most vulnerable to extreme weather events, including higher temperatures, bigger storms, variability in rain patterns, and drought. By some estimates, climate impacts could cost close to 15 percent of the region's gross domestic product (GDP) annually, and the impacts are already taking a toll. Consecutive years of drought from 2014 to 2016 left millions of people hungry. The drought in 2014 impacted 70 percent of Guatemala's land, affecting half of the country's population. In 2014 and 2015, drought

FIGURE 3.2. In Veracruz, Mexico, Central American migrants and asylum seekers cling to "La Bestia," or "The Beast," during their perilous journey north to reach the U.S. border.

Source: Joseph Sorrentino / Shutterstock.com.

destroyed $100 million in crops in El Salvador, causing economic hardship to 100,000 farmers. Migration from Central America nearly doubled between 2015 and 2016 according to the World Food Programme.[22] In 2016, the loss of agricultural production caused cases of childhood malnutrition to soar and left over 3.5 million people in the Northern Triangle in need of humanitarian aid. Hurricanes pounded the region in 2020, starting with the first named storm in the Pacific hurricane season, Amanda. In November of that year, two back-to-back Category 4 hurricanes slammed Central America again. The region depends on agriculture, especially coffee cultivation, and climate change wreaks havoc on coffee crops. Heavy rainfall can drench crops. Drought can wither them. Rising temperatures can affect the quality of the beans, similar to what happens to grapes for winemaking. Impacts of climate change, such as more rain and higher temperatures, can also encourage the growth of coffee rust, a fungus that forms on plants and can cause heavy crop losses. Bigger storms can carry fungus spores longer distances, which happened in Guatemala in 2011.

Once the rust strikes coffee plots, farmers must cut the plants down to stumps and allow them to regrow, a process that can take three years. From 2012 to 2014, the region suffered an epidemic of coffee rust in what has become known as the "Big Rust." Coffee production dropped 16 percent across all of Central America compared to the year before, with Honduras experiencing a 23 percent decline. Then between 2013 and 2014, El Salvador's production plunged from 1.7 million 60-kilogram bags of coffee to just 499,000 bags. In 2018 and 2019, droughts and floods further decimated coffee crops in the Northern Triangle.

As a result of climate change impacts, it is estimated that the land suitable for producing coffee in Central America could decrease by more than 40 percent by 2050. Many other crops will also suffer, and people will find it harder and harder to make a living as the economic prospects in Central America decline. With extreme events occurring with greater frequency, further breakdowns in economic productivity, governance, and security become more likely, giving even stronger footholds to criminal networks, including gangs. The area already suffers from some of the highest rates of gang violence in the world.

In July 2020, the *New York Times* joined together with ProPublica and the Pulitzer Center in "an effort to model, for the first time, how people will move across borders" due to climate change. Focusing on Central America, the model projects that under the most extreme climate scenarios with carbon emissions continuing to rise without mitigation, over 30 million migrants would travel toward the U.S. border by 2050. Climate change would act as the primary factor driving the migration of approximately 5 percent of these, or 1.5 million people. Even if governments around the world take "modest" action to reduce emissions, the model estimates that 680,000 climate migrants from Central America will seek entry into the United States by 2050.[23]

Similar migrations will increasingly take place all around the globe, both across international borders and within countries. Overall, the World Bank estimates that in three of the most vulnerable regions of the world, Latin America, South Asia, and sub-Saharan Africa, 143 million people will migrate within their countries by 2050 from slow-onset climate-worsened disasters like drought. That's on top of the nearly 24 million people displaced every year since 2009 by natural disasters. The policy responses to both climate change and potential migration will have significant effects on the movement of people across borders. If no action is taken to reduce climate change, more and more people will pull up stakes and attempt to cross borders in search of better lives for themselves and their families.

Alexander Betts, professor of forced migration and international affairs at the University of Oxford in Great Britain, has coined the term "survival migrants" to describe people forced to move from their countries because of an existential threat like climate change. Survival migrants could be suffering from starvation, livelihood collapse, or generalized violence. They are not economic migrants, but neither do they fit the legal definition of a refugee under the 1951 UN Convention on the Status of Refugees—someone fleeing persecution by the state. With climate change, the number of survival migrants will rise. Yet the world is unprepared for this burgeoning crisis.

In 2018, the UN General Assembly passed the Global Compact on Refugees, a new international framework. Signed by 164 nations (but not the United States), it urged countries to make plans to prevent climate-caused relocation and to support the people who have to relocate. The nonbinding document, however, accomplished little in resolving the global problem and failed to redefine the right to asylum. While New Zealand, Australia, and the United States have reached bilateral agreements with small island states to allow for resettlement when rising seas submerge those islands, the global community will need to do much more.

So what is the solution to this human crisis? How will the world handle all the survival migrants that climate change will create? Who manages, for example, the risk that climate change will render densely populated parts of the globe uninhabitable?

No nation or group of nations has yet adopted policies that fully answer these questions.[24]

The most immediate solutions for curbing catastrophe-driven displacement center on preventing people from migrating at all. Six months into the pandemic outbreak, former secretary of the U.S. Treasury Robert Rubin and former British foreign secretary David Miliband wrote an opinion piece for the *New York*

Times. The headline ran, "Borders Won't Protect Your Country from Coronavirus."[25] They argued that by helping poorer countries cope better with the ravages of the pandemic, richer nations would fulfill a moral imperative—while simultaneously serving the wealthier countries' own interests. Lending a helping hand, they argued, could better insulate the global economy against the spread of cascading economic troubles that occur as pandemic outbreaks repeatedly disrupt supply chains. Further, the unchecked spread of the disease could disrupt social and political stability and exacerbate tensions while also pushing tens of millions into deeper poverty.

The same strategy holds for coping with the impacts of climate change. Helping poorer countries build resilience would encourage people to stay safely at home, making migration a less compelling option in the first place. When the two back-to-back hurricanes pummeled Central America in late 2020, the region was already crippled by the spread of COVID-19. The ferocious storms drowned huge swaths of cropland, destroyed tens of thousands of homes, and decimated roads and bridges, upending the lives of more than five million people. As he pleaded for aid, the president of Guatemala warned, "If we don't want to see hordes of Central Americans looking to go to countries with a better quality of life, we have to create walls of prosperity in Central America."[26]

Overall, it would be less expensive to support people in their home countries than to support migrants on the move and then at their new destinations.[27] If countries respond by heightening border control and police activity to block survival migrants, those strategies will be useless for addressing the underlying patterns driving that migration, including climate change. In the face of such measures, migration will go underground with more and more people trafficked across borders by criminal networks.

For the survival migrants forced to leave their countries, the international community should define new legal protections that are recognized universally. One man solved this problem in part once before. Fridtjof Nansen, a Norwegian polar explorer, made the first traverse of the interior of Greenland on cross-country skis in 1888. After spending years researching and adventuring in the borderless Arctic, he became a diplomat and in 1921 was appointed as the High Commissioner for Refugees for the newly created League of Nations. World War I had caused nations to redraw their borders, often along ethnic lines. As a result, hundreds of thousands of people fled their homes and thus were left stateless and without passports.

As Nansen confronted the crushing humanitarian crisis, he sought to remove the barriers that international borders posed for the stateless. He created an identity document that would allow entry, travel, and work in other nations—the Nansen passport (Figure 3.3). From 1922 to 1938 about 450,000 people received Nansen passports, including ballerina Anna Pavlova, writer Vladimir Nabokov, and composer Igor Stravinsky. Both Nansen and the Nansen International Office for Refugees were awarded Nobel Peace Prizes for their work creating and issuing the passports. Despite the promise of the concept for at least allowing stateless people to travel and work internationally, the world seems far from adopting a version of the Nansen passport for those displaced or left stateless by climate impacts, much less a way for them to find permanent homes in another country.

In the absence of a current international resolution, survival migrants could end up spending decades in makeshift camps

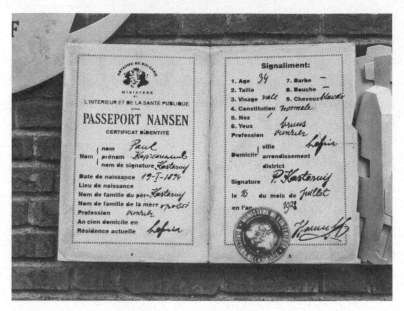

FIGURE 3.3. Nansen passport, a 1921 international refugee travel document. *Source:* Grethe Ulgjell / Alamy Stock Photo.

across the globe, waiting for Godot. As former UN humanitarian aid worker Michael Soussan wrote, "Refugees need and deserve real passports.... The world body must live up to the challenge."[28] To solve these challenges, the world needs more multilateralism, not less.

More multilateralism, not less

In his book about the 1918 Spanish flu pandemic, *The Great Influenza*, author John Barry concluded that "closing borders would be of no benefit" in fighting pandemics.[29] That conclusion largely held true in 2020 as the novel coronavirus pandemic washed across the planet, but for some countries, like Taiwan, New Zealand, and Australia, drawing up their drawbridges and acting early did lead to success. With climate change, closing borders to climate impacts is impossible. Climate change doesn't care what borders people may try to shelter behind—the impacts will keep arriving.

The pandemic has laid bare the interdependence of nations. It has further revealed how rapidly a catastrophic risk that emerges in one country can cascade, disrupting the economy and security of virtually every nation. But it's not just pandemics that threaten to cross borders. No nation can, in isolation, contain the spread of diseases from animals to humans, prevent the loss of biodiversity, or curtail the continued atmospheric accumulation of greenhouse gas emissions. Addressing these risks requires more international cooperation, not less. And to provide the necessary space to achieve cooperation, countries need multilateral institutions.

The victors of World War II knew that. The war gave the world a collective "no more" moment. The United States used that moment to prod the victorious nations to go on a peace-building spree—constructing many of the multilateral institutions existing today. Within the space of just a few years, the World Bank and the International Monetary Fund (1944), the United Nations (1945), and the World Health Organization (WHO) (1948) were launched. In the seven-plus decades since, multilateral institutions have sometimes had remarkable success, like the universal adoption of the Montreal Protocol of 1987, binding nations to phase out the production and use of ozone-depleting substances. On the other hand, when member states withhold cooperation and resources, multilateral institutions have stumbled badly, as the WHO did during the pandemic when it hesitated to criticize China for its flawed compliance with international health regulations. China's failure to notify the WHO of the outbreak or to share biological samples in a timely manner likely contributed to the spread of the virus. The WHO treated China's early statement on the nature and spread of the disease as credible and dallied over the declaration of a public health emergency, which, in turn, slowed imposition of global containment measures.

No one who built the post–World War II scaffolding for multilateralism is alive today, but their vision should remain with us. In a world of growing interdependence, multilateral institutions allow nations to find collaborative, peaceful solutions. They can provide the structure to facilitate the transparent flow of cross-border information, aid, and finance. The pandemic has proven that the current suite of multilateral institutions will struggle to deal with problems like biological threats and climate change that quickly transform from health or environmental issues into a crisis for the global community, cutting across all sectors.

The prospect of geoengineering is rapidly advancing as a possible solution to address the climate crisis. Technology is now available for any person or country with the financial wherewithal to alter the earth's atmosphere. Since 2000, geoengineering has moved progressively into the realm of mainstream science, with proposals under active discussion for spewing sulfate particles into the atmosphere and building giant mirrors to reflect sunlight, seeding clouds with chemicals to induce rain, and burying harmful gases underground. Deployment of any of these technologies could inadvertently damage another nation or community. What is the global mechanism to make sure that such consequential interventions do not run amok, causing unintended borderless disasters like destroying the soil fertility in a neighboring country or changing the rhythm of the monsoons?

As it turns out, unfortunately, none yet exists.

This was brought home to me by Dr. Leslie Field, an electrical engineer by training and founder of Ice911, an NGO focused on slowing the melting of Arctic ice caused by climate change. Her project involves spreading hollow silica microspheres, a kind of reflective sand, onto ice to reduce the amount of sunlight the ice absorbs. This, in turn, slows the rate of melting. Field describes her proposed climate intervention as "soft geoengineering," since she believes it will inflict less damage and can be more easily reversed than other interventions currently being researched. As we sat outside on Stanford University's bucolic campus in Northern California, she told me that she had already begun to raise money to fund the project, but, before the project got too far along, she wanted to find some international governmental or environmental authority to review and bless it. As I listened to her, it occurred to me that other researchers and innovators, as well as governments, might not be so conscientious. Rather, for them it might prove tempting to exploit this gaping emerging hole in international governance, which could pose untold risks for humanity. As Henry Houghton, head of the meteorology department at the Massachusetts Institute of Technology, predicted in the 1950s, the "international control of weather modification will be as essential to the safety of the world as control of nuclear energy is now."[30]

Borderless disasters already stress the current multilateral governance structure, which assigns each global issue to a specialized organization, for example, development to the World Bank, education to UNESCO, and food to the Food and Agriculture Organization. To govern an interconnected planet, this fragmented and dated approach requires a refresh, one that can convene global leaders to consider challenges such as health security, livelihood security, and resource security holistically. In addition, to thrive as intended, multilateral institutions will need even greater support, both financially and politically, from the world's most powerful nations.

Given that the pandemic has invaded every country in the world, the international community may have a "no more" moment to overhaul its aging multilateral framework. International leaders should take advantage of the crisis to put the necessary structures in place to better handle the crises of the next seven decades. Multilateralism, after all, remains our best hope for stopping borderless disasters.

CHAPTER 4

WEAVE TIGHTER
SAFETY NETS

Plan with us, not for us.

—*Marcie Roth, executive director, World Institute of*
Disabilities

I had wanderlust as a college student. In my junior year, I left for Surabaya, Indonesia, a city on the eastern tip of the island of Java, to teach English at a local university. My home for the next six months was with an Indonesian obstetrician and his family. The family lived in a compound built by Dutch colonialists. Constructed around a lush central courtyard filled with a dozen lanky palm trees surrounded by towering ferns, the compound was large enough to also house two families plus several nurses. One wing held a maternity ward. The thick walls and ceiling fans kept the buildings cool, while the courtyard's green canopy offered shade from the daytime heat. Set well above the street, the compound also avoided the flooding that accompanied the monsoon season, sometimes filling roadways with knee-high waters brimming with raw sewage that spilled from open ditches.

Across the street from the compound in a park no bigger than a tennis court sat a broken-down, wheel-less bus. It took several weeks of my walking past before I realized that three families considered the bus home. One lived on the top, vulnerable to both the relentless heat and monsoon rains. Another family lived inside the bus, baking in the sun. A third lived underneath the bus, exposed to regular flooding.

What I saw in Surabaya is the same stratification that people and countries face today when it comes to catastrophic risk. Before COVID-19 washed across the planet, the United Nations estimated that 55 percent of the world's population or some four billion people had no access to social protection programs designed to reduce poverty and vulnerability.[1] In sub-Saharan Africa, unemployment payments go to just 3 percent of the unemployed to buffer their economic loss, compared to 12 percent in Latin America and the Caribbean, 40 percent in Europe and North America, and 50 percent in New Zealand and Australia. Differences in other types of support payments for children and older people without retirement benefits were likewise divergent according to geographic regions.

These inequities have grown starker as climate change worsens weather events like the crushing heat and floods I experienced on Java. Within countries, the impact of new extremes often falls unequally, landing hardest on the poor, women, children, older adults, people with disabilities, and the marginalized. As the climate threat grows, governments, nongovernmental organizations (NGOs), and businesses should find ways to assist poorer nations as well as more vulnerable people in buffering the damage that climate change brings.

Helping others withstand future climate disruptions means weaving tighter safety nets to improve financial stability through programs like cash transfers and insurance. At the same time, more attention on reducing gender inequality and tending to the needs of people with disabilities will result in more lives saved and better outcomes for children. Finally, nations will have to come together to help the poorest countries adapt and escape crushing debt loads that will swallow any efforts to build climate resilience. This is not a new concept—far from it. At the Earth Summit held in Rio de Janeiro in 1992, the "common but differentiated responsibilities" principle (CBDR) was enshrined in the UN Framework on Climate Change, the first international agreement to address climate change. CBDR recognizes that wealthier nations likely contributed more to the accumulation of greenhouse gas emissions since they industrialized earlier, so they have a duty to play a bigger role in solving the crisis and helping those who contributed less to its creation. In the three decades since Rio, the richer nations have left that bill largely unpaid.

Women and children first

I remember the first time I learned about the lasting legacy of stunting, the impaired growth of children from a lack of nutrition, as a widespread phenomenon. I was working in my office at the White House in 2014. As head of the Resilience Policy Directorate on the National Security Council and special assistant to President Obama, I had asked for a briefing on the impacts of climate change on food security. An expert from the State Department appeared at my door with a roll of charts and diagrams under her arm. As she spread them across the conference table, she explained that as many as 20 percent of the world's children were stunted. Then she pointed to southern Asia and sub-Saharan Africa and said that three-quarters of all stunted children lived there.

The expert explained that a lack of nutrition leads to children not reaching what would have been their normal height. It can also reduce brain function and hamper organ development. Often as stunted children age, they fall behind their peers in school performance and economic productivity. Deprivation of food in early childhood could additionally result in more violent behavior in adolescents and adults. Stunting could easily run through generations, she added. A poorly nourished girl who goes on to become a malnourished mother delivers an underweight baby who grows up to be a stunted child. When many children in a country suffer from stunting, it can have a negative effect on the country's development for years.

In the two decades since 2000, the proportion of stunted children has declined thanks to measures like better maternal nutrition and health and policies that keep girls in school. But, according to the United Nations, the pandemic threatens to reverse that trend, as it upended parts of the food system. In 2019, stunting affected 144 million children under the age of five, mostly living in sub-Saharan Africa (36 percent) and South Asia (39 percent). One of the first coping strategies for families during a crisis is to eat less. A phone survey in Nigeria of close to 2,000 respondents revealed that by April/May 2020, more than half of households had reportedly reduced food consumption during the pandemic.[2] By July, the number had grown to 69 percent.

As climate change lowers crop yields and reduces the nutritional value of food, as well as disrupts food supplies, it will likely worsen the problems of food insecurity and stunting. It's not just the loss of food, however, that diminishes children's lives. It is also the loss of education caused by climate-fueled disasters.

When disasters strike, poor households—in the absence of adequate safety nets—face difficult choices, choices that can include foregoing necessary health care and nutrition or taking children out of school. A study on the impacts of climate change on education in Ethiopia found that when children experience more droughts in their early years, the odds of completing school drop by 16 percent.[3] Similarly, children under the age of three at the height of the Ethiopian famine in 1984 were less likely to finish primary school, leading to long-term economic losses. Even when students remain in school, their exposure to higher temperatures, one of the most certain impacts of climate change, could lower their academic achievement, according to an analysis of data from 58 countries and 12,000 U.S. school districts.[4] Heat alone, it appears, directly slows the development of a nation's human capital.

Calamity can also force children to go to work rather than school. In 2005 in Guatemala, Hurricane Stan increased the likelihood of child labor by more than 7 percent in the hardest-hit areas. In developing nations, girls may drop out of school to help their mothers with daily tasks, such as collecting water as water scarcity grows. Or girls are forced into marriage earlier. In Kenya, for example, with the invasion of billions of locusts that devoured crops and trees in 2020, as well as repeated droughts, families turned to marrying off their girls in exchange for dowries of fresh camel's milk and new clothes. Once girls marry, they leave school.

Women's roles and responsibilities often put them at a disadvantage in preparing for climate change's impacts, because, in general, females have fewer resources, are less likely to leave their homes unattended, and are unlikely to migrate for shelter and work when disaster hits. Structural inequities also play a role in the poorer outcomes for women and girls. Food insecurity is higher for females. Estimates for 2016–2019 indicated that women suffered greater food insecurity in every region.[5] In some cultures, girls and women are the first to go without when food runs short, despite often serving as the main procurers of food, water, and fuel. In addition, women serve as caretakers, looking after the children, the elderly, the sick, and even the family's assets. With drought repeatedly striking Kenya in the 2000s, women have had to walk longer distances to access water, an average of close to seven miles every day (Figure 4.1). In South Asia, long power outages, which can result from infrastructure failure during climate-worsened extremes, have reduced per capita income, female participation in the labor force, and girls' time available for study.

FIGURE 4.1. Girls travel miles from home in Dar Es Salaam, Tanzania, carrying heavy loads of water above their heads, which has long-term effects on their muscular skeletal structure.

Source: Photographer RM / Shutterstock.com.

Beyond the simple needs of survival, women and girls' lower socioeconomic status gives them unequal access to information and assets that could assist during a disaster. Additionally, cultural customs for dress such as bulky, long-flowing garments or caregiving responsibilities may make it more difficult for them to evacuate. Because of social norms, they may not have been taught to swim or climb trees, placing them in greater danger if floodwaters rise. If disasters occur during pregnancy or childbirth, the vulnerability of females grows.

Simply put, when disasters strike, women are at greater economic risk and are more likely than men to die. According to the United Nations, women and children are 14 times more likely to die as a result of a natural disaster.[6] In 2008, during Cyclone Nargis in Myanmar, twice as many women aged 18 to 60 died as men. When Cyclone Bhola struck in what is now Bangladesh in 1970, killing half a million people, 14 times more females died than males. With wind speeds of 185 kilometers per hour (115 miles per hour) and a storm surge of up to six meters (20 feet), only those who could climb high enough and were strong enough to hold on survived. That meant children, women, and older people died. With expansion of early warning and other preparedness measures, the ratio of loss of life among women and men dropped during Cyclone Sidr in Bangladesh in 2007. In that storm, five times more women died than men.

As climate impacts worsen, the threats to women and children multiply. If a disaster causes pregnant women or children to suffer from a lack of nutrition or to miss out on schooling, it can have long-term negative consequences. Communities and their countries will experience reduced productivity as tens of thousands of children reach adulthood in poorer health and with lowered intelligence as well as academic achievement. To avoid these outcomes, adequate access to food remains essential, as does schooling. Ultimately, achieving gender parity will also help lessen some of the disparities.

COVID-19 illustrates what's at stake. In both developing and developed countries, the disease hit women and children particularly hard. Reports of domestic violence and child abuse exploded around the globe, prompting UN Secretary General António Guterres to urge "all governments to put women's safety first as they respond to the pandemic."[7] Reports of child marriage surged as well. Child marriage is almost universally banned, and yet prior to COVID-19 the United Nations estimated that every single day 33,000 girls are forced into marriage globally. The pandemic has caused those numbers to soar. The international NGO Save the Children estimated that the pandemic put 500,000 more girls at risk of child marriage in 2020 alone. It further predicted that as many as ten million children—mostly girls—may never return to school due to COVID-19's spread.

During the pandemic in 2020, the unemployment rate for both men and women increased. But for women, who are already less likely to be in the workforce, the unemployment rate was 1.7 times higher than for men—321 million women as compared to 182 million men in the second quarter of 2020 in 55 countries.[8] The U.S. government reported that in December 2020, as the pandemic raged, women lost 156,000 jobs while men obtained 16,000, and Latinx and Black women lost even more jobs than white women.

Both advanced and developing countries experienced disruptions or total suspensions of childhood vaccinations during the pandemic. Working women lost their jobs when childcare became unavailable. In communities where adequate nutrition was not available, children's development suffered. School closings and unequal access to quality online education reduced learning, possibly with long-term consequences for millions of students. According to a survey conducted by the International Labor Organization, 44 percent of 18- to 29-year-olds in low-income countries reported receiving no education or training in the spring of 2020, while the figure dropped to just 4 percent in high-income nations.[9]

In 2015, 193 nations came together to agree on a plan "to achieve a better and more sustainable future for all" by 2030.[10] The plan enshrines 17 Sustainable Development Goals (SDGs). The very first goal is eliminating poverty. The second goal is "zero hunger." The fifth goal is gender equality. Despite the fanfare surrounding the SDGs, progress on the goals foundered in the plan's first five years. Then the arrival of the COVID-19 pandemic shook the whole effort to "its very core," causing the first increase in decades in global poverty.[11]

As countries strive to recover from the pandemic's devastation, they should refocus on these essential goals. Doing so will leave them better prepared for the ravages of climate change. Notably, a few countries used the pandemic to double down. Canada carved out some benefits to specifically target women, while other countries directed funding to sectors where women were heavily represented. Burkina Faso waived fees for fruit and vegetable sellers, and Argentina provided masks to home-based workers. All of this is a good start, but the efforts to reduce hunger and gender inequality need to gain momentum. The pandemic has given the world a glimpse of the lasting penalties that will mount in the face of escalating climate risk.

"Plan with us, not for us"

It's not just women and children who deserve special attention when it comes to dealing with climate change. It's also people with disabilities. The United Nations estimates that worldwide there are one billion individuals living with disabilities. People with disabilities and older adults are two to four times more likely to die in a disaster. During the first weeks of the pandemic, over 7,000 residents of nursing homes in the United States died. The facilities became "death pits."[12] By November 2020, the number of deaths had grown tenfold to more than 70,000. People in long-term care facilities accounted for about 6 percent of all coronavirus cases and 35 percent of all deaths from the disease in the United States, even though less than 1 percent of America's population lives in such facilities.

Research has shown that in the aftermath of hurricanes, including the hurricanes that hit the United States in the years leading up to the 2020 pandemic, nursing home residents and those relying on life-sustaining medical equipment also suffered disproportionately.[13] When utilities fail, electricity-dependent medical services also fail. New York City's treatment of Coney Island residents illustrates these risks.

Coney Island is one of New York City's poorest neighborhoods. When Superstorm Sandy struck in 2012, one in seven Coney Islanders was unemployed. According to census data, the average household earned $31,000 a year, and one in four residents fell below the poverty line.[14] Many Coney Islanders who were older and/or suffered from disabilities lived in high-rise public housing units. After the storm knocked out power, they found themselves stranded without electricity, heat, or elevator service, "effectively held hostage on the upper floors of their apartment buildings."[15] Temperatures dropped to near freezing, and still there was no power, heat, or hot water. The city housing authority, overwhelmed by the disaster, took weeks to even assess the status of the Coney Island residents (Figure 4.2). One of the wealthiest cities in the world did not provide assistance when some of its most vulnerable residents needed it most.

What happened in Coney Island concurs with findings by the UN Office for Disaster Risk Reduction that "disaster discriminates on the same lines that societies discriminate against people."[16] Marcie Roth understands this firsthand. I first met Roth in 2016, when she headed U.S. federal emergency efforts to assist people with disabilities. In 2020, she became executive director of the World

FIGURE 4.2. Hurricane Sandy flooded streets, destroyed buildings, and stranded thousands of Coney Island residents without light, heat, or water for weeks.

Source: John Huntington / Shutterstock.com.

Institute on Disability (WID), the oldest organization continually led by people with disabilities. Active in the disability rights movement since high school, she herself has a disability, raised two children with disabilities, and married a man with a disability.

Roth believes that what makes people with disabilities vulnerable in disasters "is the failure of communities and governments to plan for the inclusion of people with disabilities in every aspect of the disaster cycle, including community preparedness and disaster exercises, accessible alerts and warnings, building and community evacuation, sheltering and temporary housing, access to health maintenance and medical services, and all aspects of the recovery process," as she told the U.S. Congress.[17]

Older people, who may suffer from disabilities beyond simply age-related conditions, fall victim to disasters at astonishing rates. Reduced mobility can make it impossible for them to reach safety. More than 70 percent of the fatalities from Hurricane Katrina in the United States in 2005 were among people 60 and older. During the 2010 Japanese earthquake and tsunami, over 65 percent of the fatalities were also among people 60 and older, and more than 90 percent of those died by drowning. Roth believes that the pandemic has been "disproportionately horrific for people with disabilities," including the aged.

At WID's helm, Roth has turned the organization's focus to disaster preparedness. In her words, disaster efforts should "plan with us, not for us." By way of example, she asks, "What about this practice of putting houses 12 feet off the ground to avoid climate-worsened flooding? What does that mean for the housebound or those that cannot walk—both for the ability to socialize and to evacuate during a hurricane?" Roth does not hold back, not with me, nor when she testified to Congress that American communities remain unprepared for disasters. She argues that communities should plan together to find inclusive solutions as they confront increasing calamities.

Roth is right. She cites the UN secretary-general's Disability Inclusion Strategy launched in 2019 as one way to begin to tackle the problem internationally. It requires the United Nations to begin to make itself and its programs more accessible. Until the voices of people with disabilities are included in all levels of planning, they will remain at much greater risk to the disasters fueled by climate change.

As policymakers grapple with the unique risks climate change poses to women, children, older people, people with disabilities, and the marginalized, they need to find ways to keep these and other vulnerable populations from slipping deeper into poverty. That means weaving stronger safety nets to catch people before the slide becomes too steep for recovery. The pandemic has given nations a chance to weave quickly. Countries need to take that work and tighten it to provide the kind of social protection that climate change requires.

Tightening safety nets

Hurricane Mitch struck Central America in 1998, killing over 11,000 people. Only the Great Hurricane of 1780 has caused more deaths in the Atlantic region in recorded history. A Category 5 storm, Mitch dumped record amounts of rain in just four days— 190 centimeters' (75 inches') worth—across the Northern Triangle of Central America (Honduras, Guatemala, and El Salvador). It inflicted the greatest pain on Honduras, the second-poorest nation in the Western Hemisphere at the time. The torrential rains caused rivers to overflow and mudslides to bury villages. So many roads and bridges washed away that existing road maps became useless. More than 7,000 Hondurans were killed, and 1.5 million people, about 20 percent of the country's population, were left homeless. In 1999, the World Meteorological Organization retired Mitch as a name for hurricanes in recognition of the extraordinary damage and loss of life.

The then-Honduran president lamented that the storm had cost the nation half a century's worth of development. "Our agriculture is in shambles. All of our major crops, our export products ... gone," he told CNN.[18] Agricultural losses amounted to close to $3.8 billion, about 70 percent of Honduras's gross domestic product (GDP). The storm hit banana production particularly hard, ruining 80 percent of the crop. Twenty-five thousand banana workers lost their jobs.

Hurricane Mitch pushed millions of people deeper into poverty, making an already poor country suddenly even poorer. Moreover, those who still had homes had to leave in search of work because their farms could no longer sustain them. Approximately 150,000 people left for the United States within six months of the disaster. In 1999, the U.S. government gave 57,000 Hondurans permission to stay in the United States temporarily in light of Mitch's utter destruction.

Two decades later, Honduras has yet to recover, remaining one of the poorest countries in the Western Hemisphere with nearly two-thirds of its population in poverty. Its government institutions remain weak, with corruption widespread among police and government officials, and drug lords and gangs controlling swaths of territory. Honduras is also one of the nations in the world most vulnerable to climate-related hazards. In the midst of the pandemic, two back-to-back Category 4 storms pummeled the nation. In scenes rivalling the destruction of Hurricane Mitch, the storms left thousands homeless, flattened infrastructure, and washed away over 50,000 hectares of cropland. To make matters worse, scientists predict that by 2050 western Honduras could become a climate "hotspot," with soaring temperatures and increasingly severe droughts as rainfall decreases by up to 20 percent.

Disasters punch poorer countries like Honduras harder, because, as the World Bank has determined, they have less protective infrastructure to buffer them from extreme weather events. Take floods, already the most common natural disaster and one that will worsen with climate change. Poor countries have fewer flood protection structures in place. They also have more people living in low-lying flood areas that lack adequate drainage or on steep hillsides subject to mudslides. Sub-Saharan Africa is a standout for poverty and flood risk. Although the area houses about 10 percent of the global population at risk of flood, it has more than 50 percent of the global poor facing high flood risk, according to the World Bank.[19]

Persistent droughts, storm surges washing saltwater over fields, and urban flooding can bring financial ruin to the poor, slowly or suddenly. Disasters sweep away or force the sale of assets such as fishing boats, market stalls, livestock, and land, depriving people of their only means for making a living. Disasters take away even the most meager of possessions—mud houses with scrap tin roofs, a few sacks of grain for lean times, or the goat that provided milk. They put people at greater risk for water-borne diseases as floodwaters spread, and respiratory illnesses spike as wildfires rage. With changing weather conditions, insect-borne diseases like malaria and dengue expand their geographical range.

Disasters also kill at a much higher rate in poorer nations, at the rate of 130 people for every million, compared to 18 per million in richer countries.[20] Over 90 percent of reported disaster-caused deaths occur in poorer countries. The economic damages from disasters as a percentage of GDP are much higher in poor countries. The United Nations has determined that of the ten worst disasters with the highest economic losses as a percentage of GDP between 1998 and 2017, only one was in a high-income area, the U.S. territory of Puerto Rico, after it was struck by Hurricane Maria in 2017.[21] In a particularly crushing set of blows that same year, Hurricane Maria's ferocity also cost the tiny Caribbean country of Dominica 260 percent of its GDP, while only two years before, Tropical Storm Erika had inflicted damage worth 90 percent of that island nation's GDP.

So how can nations break this cycle of increasing vulnerability and inequality to become more resilient? Weaving tighter safety nets can lead to better outcomes. Tightening safety nets can prevent a plunge into poverty and build future resilience. Two essential ways to provide assistance are through cash transfers and access to insurance.

Cash infusions can drive resilience

Before 2020, fewer than half the world's nations had legislated the provision of unemployment insurance and less than a quarter of unemployed people received benefits. But then the pandemic hit. The spread of the disease drove an economic contraction that pushed over 100 million people into extreme poverty where they were living on less than $1.90 a day. Sub-Saharan Africa and South

Asia, the regions where most of the global poor are concentrated, were the most affected.

Across the planet, the pandemic has exacerbated income inequality, with those already at the low end of income scales hurt the most by supply chain disruptions and lockdowns. Increasingly, it's the urban poor at greatest risk of slipping from poverty into extreme poverty, rather than rural populations, which have historically been at the greatest risk. The World Bank believes the convergence of climate change impacts with the unfolding of the pandemic, as well as added pressures from global conflict, threatens to reverse a two-decades-long trend of poverty reduction.

What modest gains the globe had made in reducing poverty all but vanished as the pandemic trampled livelihoods. Governments around the world responded to the crisis by rapidly expanding and bolstering safety nets. For example, Afghanistan designed a relief package aimed at both the urban and rural poor, covering about 90 percent of all households. Rural households received approximately $50 worth of food and hygiene products, while urban residents received the equivalent of $100 in cash and in-kind support. Similarly, Canada, which had toyed with the idea of creating a long-term basic income program in the 1970s, went into high gear, providing over $60 billion in pandemic emergency benefits through direct bank deposits to citizens. Included among the beneficiaries were the self-employed, child caretakers, and quarantining households. By mid-2020, 200 countries or territories had put in place some type of social protection and more than 50 of these involved the expansion of cash transfer systems.

Even before the pandemic, cash transfers had proven effective in braking poor families' descent into penury. Brazil pioneered the idea of conditional cash transfers to reduce poverty in the 1990s.

The Bolsa Familia program targeted poverty by providing cash on the condition that recipients take certain defined actions to improve their lives, like enrolling children in school and receiving vaccinations. Once empirical studies proved the efficacy of Brazil's program, a "quiet revolution" took place.[22] Developing countries around the globe began to invest in cash transfer programs.

Research has shown that the benefits of post-disaster transfers of cash and other goods far outweigh the costs. In a study of 117 countries, post-disaster transfers yielded a cost-benefit ratio of one to three, and in 11 countries, every $1 spent yielded more than $4 in well-being benefits.[23] Cash transfer programs have the potential to improve resilience to climate shocks. Tanzania's experience provides proof.

Tanzania's Productive Social Safety Net (PSSN) program is geared toward the extremely poor, linking cash payments to participation in public works projects when the demand for labor in agriculture is low. In 2019, the Tanzanian government, in collaboration with the World Bank, approved a new phase of the PSSN program, aimed at helping extremely poor households transition to more sustainable livelihoods. The public works aspect of the program prioritized funding projects offering the highest climate resilience potential, such as the creation of water catchments and dams to expand access to water for both farmland irrigation and drinking water for livestock.

Cash transfers can reduce the need for food aid and allow the poor to accumulate assets that could help them deal with new weather extremes. For example, in Niger, cash transfers have helped build resilience to climate shocks by promoting savings that can buffer disaster losses, diversifying income sources to make recovery easier, and providing a cushion against the pressure to make a poor coping decision, such as selling livestock. Similarly, requiring recipients to acquire new skills can prompt people to try new measures that enhance their resilience, like using drought-resistant seeds or improving irrigation.

As the planet suffers ever-greater impacts from climate change, developing countries, in collaboration with international lending institutions and NGOs, should use "muscle memory" built from the pandemic experience to strengthen and reinforce cash transfer programs. The rapid rollout of assistance in the face of spreading disease proves that governments can expand transfer programs quickly—either by providing greater amounts of money or expanding the pool of beneficiaries, or both. Existing programs have shown that quick action cuts losses. Countries should continue to refine cash transfer approaches so that they are easily scalable in times of trouble. Doing so will give countries a valuable tool as they seek greater resilience to climate shocks. Insurance is another necessary tool for the resilience toolbox.

Closing the insurance gap

Without insurance, a single crisis within a family, like the death of a breadwinner or a flood or locusts that destroy crops, can mean impoverishment that lasts for years. For a nation, a disastrous storm or drought can constrain development and retard economic growth. If insurance has not "penetrated" a country to a significant degree, individuals, governments, and businesses have to pay for the damages themselves, making bouncing back that much harder and less likely. Governments in places where insurance is lacking also often snare funds designated for other purposes to make post-disaster repairs, which, in turn, slows long-term economic development. In contrast, having insurance means that people, businesses, and governments have money available to pay for the loss of infrastructure, shelter, livelihoods, and assets like equipment and livestock, making reconstruction and disaster recovery happen that much faster. In other words, by providing a safety net for the ravages that catastrophes can bring, insurance helps protect against poverty and, thus, build resilience.

In poorer countries, the "protection gap," the difference between insured losses and uninsured losses, is huge. As of 2020, over 90 percent of losses in developing nations are uninsured, a figure that has remained steady for almost half a century. In very low-income countries, insurance coverage often falls below 1 percent. For instance, the insurance rate for Madagascar, one of the poorest countries, is less than two-tenths of 1 percent. Bangladesh, a country highly threatened by natural disasters, is also among the most underinsured countries in the world. Insurance covers just 9 percent of the economic losses caused by natural hazards in Asia. Only 5 percent of commercial and industrial property and only 1 percent of homes are insured in China, according to most estimates. In contrast, in North America natural catastrophe coverage climbs to 40 percent.

Climate change threatens to widen this protection gap even further, because insurance will become more costly with increasing risk. The urgency of this problem has led to new efforts to try to close the gap. One study found that just a 1 percent expansion in insurance penetration would lead to a 22 percent reduction in the disaster burden on taxpayers.[24] As we saw in Chapter 3, some countries have banded together to create regional risk pools to obtain cheaper insurance. Others have created compulsory nationwide insurance schemes to help their populations bounce back quickly from specific risks or multiple perils.

Consider Morocco. In 2019, with the help of the World Bank, Morocco created the Solidarity Fund against Catastrophic Events (Fonds de Solidarité contre les Événements Catastrophiques [FSEC]). The FSEC provides partial compensation to uninsured households that suffer injury or loss of their residence as a result of catastrophic events. Designed to cover 95 percent of the population, the program will provide insurance to five million people living on less than $3.10 a day. One way it generates revenue is by placing a 1 percent tax on insurance premiums.

Expansion of microinsurance, low-cost insurance intended to protect low-income people, also holds promise for building resilience. The Philippines is among the most vulnerable countries in the world to natural disasters, especially typhoons. An average of 20 typhoons pass through the nation each year. Since 2010, banks in the Philippines have sold microinsurance policies designed to protect people living on less than $4 a day from damage caused by disasters. The arrival of Super-Typhoon Haiyan in the Philippines in 2012 made clear the wisdom of that decision. Carrying the highest wind speed ever recorded on land up to that time, 313 kilometers (195 miles) per hour, the storm displaced millions, many of them poor, and caused more than $700 million in crop and infrastructure damage.

But microinsurance sped recovery for many Filipinos. Insurers paid over 100,000 microinsurance claims, almost all for disaster coverage. The average payout was $108, which recipients typically spent on either house repairs or restarting their livelihoods. In the wake of the storm, insurance demand skyrocketed, expanding coverage to about 20 million properties and lives, leaving the

Philippines with one of the highest rates of insurance penetration in Asia.

Microinsurance is usually a much cheaper product than traditional indemnity insurance. The payouts are smaller, often relying on a parametric trigger—the occurrence of a verifiable fact that triggers payment. As we saw in Chapter 3, nations have used parametric insurance to help prepare communities in advance of disasters as well as to help recover from them afterward. Without the complexity of having to conduct a loss investigation, parametric insurance claims are paid quickly. Microinsurance also has the advantage that it can cover more people for a variety of risks ranging from accidents to loss of crops, health, and property. Policymakers can tailor products to meet the needs of groups that have fallen outside of traditional insurance markets, such as farmers in rural Africa.

In the Horn of Africa, severe drought causes 75 percent of livestock deaths, often leaving herders with no livelihood. The Kenya Livestock Insurance Program (KLIP) provides parametric insurance payouts to pastoralists based on satellite imagery of local vegetation. A change in the color of the ground in the images from green to yellow means that drought conditions have arrived and livestock can no longer survive simply by grazing. When the yellow image appears, the program provides money, which the pastoralists can then use to purchase supplemental feed and water.

The program serves as preventative protection by saving animals and sustaining livelihoods when it covers the farmers' "almost losses."[25] The Kenyan government assists by providing subsidies for the insurance premiums. As of 2019, the KLIP had helped 32,000 pastoralists in Kenya. The general idea holds promise for cattle farmers in drought-prone countries and regions around the globe, including the Middle East, parts of China, and other areas in the Horn of Africa.

Microinsurance can also drive greater investment in resilience measures. A study in Northern Ghana showed that, until farmers had access to microinsurance, they resisted agricultural improvements for fear of suffering a loss.[26] But when the farmers purchased parametric insurance, it increased their willingness to invest in things that could bolster output, like fertilizer, and spurred their ambition to invest in cultivating more land and putting more effort into soil preparation. By reducing the perceived risk, farmers felt they could also experiment with new, more productive varieties of crops. Bundling insurance policies with other products can yield even greater resilience dividends. For example, access to insurance allows farmers to use their livestock and crops as collateral to obtain loans to buy better equipment or more seed, thereby increasing not only their output but also food security.

Conditioning microinsurance on efforts to reduce risk can likewise build resilience. Consider the R4 Rural Resilience Initiative, a program pioneered in 2011 by the World Food Programme in partnership with Oxfam. R4 insures even the poorest farmers in Africa. In exchange for an agreement to invest in resilience measures like building dikes and rehabilitating low-lying lands for rainfed agriculture, the farmers get insurance. Similar efforts have caused farmers in Ethiopia to invest more in fertilizer and assets like farm animals and encouraged farmers in India to stop burning fields to prepare them for planting, a common practice that can poison the air for hundreds of miles. Over time, these programs help farmers reduce their vulnerability to disasters.

Closing the insurance protection gap in poorer nations has to remain at the top of the to-do list for international development agencies and NGOs. Insurance helps provide the money to jumpstart resilience by both driving better predisaster choices and fueling a rapid recovery after disaster strikes. But it's not just poorer countries that could face a widening insurance protection gap as a result of worsening climate impacts. Richer nations could too.

"A 3- or 4-degree world is not insurable"

In 2016, I met with the CEO of a major reinsurance company. As we sat in his sleek office overlooking the Connecticut River in Hartford, Connecticut, he expressed his concern that climate risk would grow too large for even the reinsurers, the insurance companies that insure insurance companies against catastrophic risk. He then added, "And you know what that means, don't you?" He answered the question before I could: "It means that governments will have to pay for the damage, if they can." In other words, as private insurance companies stop offering insurance in the face of escalating risks, richer countries could see insurance coverage shrink.

Catastrophic risks have caused private insurance companies to exit insurance markets before. After Hurricane Andrew inflicted unprecedented losses in Florida, the state strengthened both its building codes for wind and insurance oversight to keep insurers from fleeing the state and/or raising premiums sky high. A similar story unfolded in Australia in 2012. The Suncorp Group, one of the largest general insurers in Australia, stopped offering and renewing policies in an area that had suffered three floods in two years. Like Florida, the Australian government sprang into action.

Only after it invested in the construction of levees to reduce the risk of future flooding did the insurers start offering policies again.

The risk remains that private insurance companies may choose not to do business in the face of more extreme climate-fueled disasters. That could leave property owners without insurance, which would likely depress real estate prices while adding pressure on the government to pay a greater share of the damages. This problem has begun to emerge in California, the fourth largest insurance market in the world. After historic fires in 2017 and 2018, some insurance companies stopped writing new policies and renewals. In response, the state mandated that insurance companies could not drop homeowners who lived near the burned areas for the following year. But then even bigger wildfires rampaged the state in 2020, causing the California Department of Insurance to extend its order for another year. As rising temperatures fuel more destructive wildfires, many Californians may discover that fire insurance is unaffordable and increasingly difficult to obtain.

Governments will face pressure to fill the resulting insurance void. Should they choose to do so, however, it could prove politically impossible to extract themselves in the future. What happened in the United States with flood insurance provides a cautionary tale. When private insurers balked at providing flood insurance after a series of devastating floods in the United States in the 1960s, the U.S. government created the National Flood Insurance Program (NFIP). Judging by the NFIP's performance in its first 50 years of existence, government efforts to provide catastrophic risk insurance can stumble spectacularly.

The NFIP is widely viewed by insurance experts and government watchdogs as broke and broken. Its payouts to homeowners far exceed the premiums it receives, while also providing perverse incentives to rebuild at-risk homes (sometimes over and over) and build new homes in flood-prone areas. As of 2020, the program was over $20 billion in debt with $16 billion previously forgiven. Yet despite the NFIP's many flaws, politicians' repair efforts have failed time and time again. Proposed reforms, such as raising rates or refusing insurance for properties that suffer repeated flooding, generate too much opposition from homeowners, local communities, and developers, forcing the program to go deeper and deeper into debt.

If solutions are not found, private insurers, as the reinsurance CEO predicted, could altogether stop offering certain kinds of insurance in the face of accelerating climate risk. As the global reinsurer AXA announced in 2018, a 3- or 4-degree-Celsius (5.4- or 7.2-degree-Fahrenheit) world—the average temperature increase that the planet could experience by the end of the century if we continue on the current course of greenhouse gas emissions—is "not insurable."[27] Should private insurers stop insuring climate-related risks, it will become ever more difficult for the world to recover, adding to the misery of all but especially the most vulnerable.

Getting out from under debts so that all can thrive

One of the deeply troubling aspects of climate change is that those least responsible for causing global warming will pay the heaviest price. Take Bangladesh, with its pancake-flat terrain crisscrossed with rivers and high vulnerability to sea-level rise. The country is a poster child for the inequitable cost that climate change inflicts. In the midst of the pandemic in 2020, Bangladesh suffered from two events consistent with climate trends, Cyclone Amphan in

May and relentless rain in July, which flooded over a quarter of the country's landmass according to government estimates. It's noteworthy that the average American bears responsibility for 33 times more carbon emissions than the average Bangladeshi.

For years in international climate negotiations, poorer countries have sought compensation from richer countries for the harm the wealthier nations have caused, or in the parlance of the United Nations, "loss and damage." Wealthy countries pushed back, concerned that they could face legal liability. In 2009, richer nations pledged to mobilize $100 billion per year by 2020 from a variety of sources to help poorer nations address climate change. The promises have not been fulfilled. Not only has the money run short, but also, from 2013 to 2018, 70 percent of the funding went to climate mitigation and only 21 percent to adaptation, with the remainder to cross-cutting projects.[28] Adding to the challenge for poorer nations is that financing has increasingly come in the form of loans, rather than grants, with three-quarters of the funding dispensed as loans in the period from 2016 to 2018.

The COVID-19 pandemic forced debt-laden nations to incur even more debt to cover tax revenue shortfalls while ramping up spending to strengthen safety nets and provide medical assistance. Analysts call the back-breaking increases a "debt tsunami."[29] Tourism-based economies have suffered the most. For example, tourist-dependent Barbados borrowed more than $350 million, increasing its debt-to-GDP ratio to above 130 percent. The International Monetary Fund (IMF) projected that global public debt would climb 19 percentage points above 2019 to more than 100 percent of GDP in 2020. In response to the burgeoning crisis, the G20, a group of the world's biggest economies, agreed to freeze loan repayments for the world's poorest nations from April 2020 until June 2021.

But that solution did not provide lasting relief to the mounting debt poorer countries face, and climate-fueled disasters will only add to their pandemic-worsened debt woes. In 2018, more than 60 countries saddled with low credit ratings could only access capital at interest rates higher than 18 percent for projects longer than two years. As financial markets price climate risk with greater frequency and accuracy, borrowing money will become even more expensive. Climate disruptions will require increased public spending on reducing risk as well as on long-term recovery efforts. Countries may find themselves unable to provide necessary government functions because their revenues must go to debt service. Given the perfect storm of climate change and the pandemic, it's time for the international community to, in the words of the pope, "renew the architecture of international finance."[30] Radical loan adjustment is called for. It's time to consider a debt jubilee.

Debt jubilees likely date back to the Bronze Age in the Sumer civilization. When new rulers took the throne, they would declare a debt amnesty to keep their people from falling into debt bondage. Rulers also sometimes declared debt jubilees in the wake of disaster, when survivors faced ruinous debt, to fund their recovery. During the pandemic, the idea has resurfaced. On Easter Sunday in 2020, Pope Francis called for a "reduction, if not the forgiveness," of debt for poor countries.[31] NGOs joined in the call for a debt jubilee for impoverished countries dealing with COVID-19.

Yet, the international financial system lacks an effective means to help nations extract themselves from the debt cycle. Reform is urgently needed. The IMF created a voluntary system in 1996—the aptly named Heavily Indebted Poor Countries

Initiative (HIPC)—with the goal of ensuring that no poor country confronted a debt burden it couldn't manage. Since then the HIPC has approved debt relief for over 35 countries, the vast majority of them in Africa. The HIPC has done admirable work, but because not all creditors participate, some countries may have a debt albatross still around their necks. That means that lenders like the World Bank are left with only moral persuasion to achieve the HIPC goal.

In recent decades, philanthropists and development agencies have created other avenues for some measure of debt relief that could also address climate concerns. One route for reform is to tie loan adjustments to progress on other issues. Consider "debt-for-nature swaps." These exchanges lower the interest payments for the debtor if it agrees to conserve natural spaces. Actor Leonardo DiCaprio helped the small island nation of Seychelles to reduce its burden when his foundation assumed some of the country's debt in exchange for the creation of two new marine parks in 2018. Swaps are particularly attractive for addressing climate change since the protected lands can act as carbon sinks while providing a buffer against worsening climate impacts. As of 2014, the United States Agency for International Development (USAID) had reached 20 debt-for-nature agreements with 14 countries, ranging from Brazil to Bangladesh.

In addition to debt-for-nature-swaps, creditor nations could consider "debt-for-climate-swaps," which would result in a partial cancellation of debt while transforming the remaining debt payments into investments in climate action. Debt swaps, however, are not trouble free. Beyond the challenge of monitoring adherence to conservation and climate practices over the long haul, the programs rely on donor funding, which, in turn, often depends on the foreign assistance budget in richer countries or the generosity of billionaires. Debt swaps also require sophisticated technical know-how to ensure a pipeline of credible investments. Further challenging success is the small amount of debt typically involved, which limits the overall effectiveness of the programs.

To address the risks posed to borrowers by growing disasters, France's development agency pioneered the concept of countercyclical loans. Countercyclical loans automatically suspend loan repayments for up to five years should a particular shock, for example, a fall in GDP or a hurricane, affect a nation's ability to repay. This grace period gives nations time to get back on their feet, at least theoretically, before they need to recommence payments. Given their vulnerability to climate-fueled storms, Caribbean nations have sought "natural disaster clauses" in their loan agreements since 2015 to ease debt requirements in the wake of disasters.

But none of these reforms gets to the heart of the problem, the unsustainable load of cumulative debts crushing poorer countries. These countries need cash to prepare for climate impacts and embrace clean energy. A transformation in current international lending practices is required. The participation of China, the world's largest bilateral lender to developing countries, could jumpstart the progress. Whether China chooses to play a leadership role or not, multilateral lending institutions and creditor nations, as well as private lenders, have to develop an approach that allows developing nations to emerge from the crises currently squeezing them.

The essential point is that the pandemic has given the world a searing glimpse of how climate extremes could multiply the threats faced by the most vulnerable. Nations should use this moment to inform new approaches for building greater resilience. The good news is that countries have shown a wide range of responses to the pandemic, providing an opportunity for nations to learn from each other as they seek to do more to protect their populations, including those at greatest risk.

CHAPTER 5

JUMPSTART RESILIENCE

We can only see a short distance ahead, but we can see
plenty there that needs to be done.

—*Alan Turing, English mathematician and computer scientist*

The state of Michigan in the United States and the province
of Ontario in Canada share a border that stretches over 700
miles (1,126 kilometers). In August and September 1986, heavy rain
fell along that border. The biggest deluge came between September
10 and September 12, when up to 14 inches (35 centimeters) poured
from the sky onto the already saturated ground and caused widespread flooding. The two areas hit by the rains in Ontario and
Michigan were comparable in population, levels of urban development, geography, and volume and frequencies of rainfall. The
plight of each area, however, differed widely.

In Michigan, the Great Flood of 1986 inundated 30,000 homes
and ruined crops. Eleven dams failed. It was the worst flood in
the state's recorded history and the worst in 50 years anywhere
in the United States. Flooding caused over half a billion dollars'
worth of damage, making it the most expensive natural disaster
in Michigan's history up to that time. But, across the border in
Ontario, local communities had a much easier time of it. Even
though the physical flooding in the province was actually greater

than in Michigan, the damage was minimal, costing only half a million dollars.

What can explain the difference? The answer lies in the presence—or absence—of land-use policies aimed at reducing flood damage.

Ontario began regulating development in its floodplains in the early 1940s. But the crucial moment occurred in 1954, when Hurricane Hazel swept across southern Ontario, dumping more than 11 inches (28 centimeters) of rain in just 48 hours. The storm and its floodwaters killed 81 people and left 1,800 families homeless. "No more," provincial leaders vowed. The destruction marked "a major turning point in political awareness and energy . . . for reducing future flood losses in Ontario," the Canadian Institute for Catastrophic Loss Reduction reported.[1]

The provincial government restricted development and rebuilding in flood-prone areas and instituted flood-proofing programs to reduce future physical damage by focusing on better drainage systems. Authorities also created the first flood risk maps to steer development away from at-risk areas and to serve as the foundation for land-use and infrastructure planning. In areas prone to flood, legislation prohibited the construction of new government buildings and restricted the use of government funds to support new development. New structures built in flood zones became ineligible for future government disaster assistance. And with insurance premiums rising in the at-risk areas, development largely moved to areas outside the floodplain.

These measures reduced Ontario's vulnerability to the excessive rains that caused severe damage in Michigan, research has shown.[2] In contrast to the province's discouragement of building in at-risk areas, the state of Michigan and the U.S. government had taken an almost completely "hands-off" approach to regulating development in the floodplain. Although Michigan had prohibited basements in flood-prone areas and instituted some elevation requirements to avoid flooding before 1986, enforcement was lax and builders had often ignored the rules.

Even more important, Michigan permitted extensive development in the floodplain. Local communities chose how much development to allow and allow they did, typically doing nothing to deter building in areas prone to flooding. The inevitable result was huge losses. In the town of Vassar, Michigan, for example, the raging Cass River, at 14 feet (4.3 meters) above flood stage, swept through every downtown business and 50 homes that had been built in the floodplain.

This story of two countries and one flood holds a crucial lesson: in the face of accelerating risk, including that from floods and wildfires made more deadly by climate change, governments and businesses should pursue policies that jumpstart resilience. Admittedly, it's a difficult task, since communities struggle to muster the political will to act without experiencing a "no more" moment like Ontario did after Hurricane Hazel. And the task is growing more challenging every year, given the acceleration of climate change, the resulting inability to plan for the future based on the climate of the past, and the ensuing large uncertainties about how the impacts of climate change will unfold.

But there are important steps decision makers can take to jumpstart future resilience. One crucial step is requiring companies to disclose climate risk, giving them a deeper understanding of what they're up against and what they might need to do to prepare. Another is using scenario analysis to test operations against different imagined futures. More generally, all investments, policies, plans, and programs should be routinely screened for climate resilience. At a minimum, to avoid wasting precious resources, any investments in long-lived infrastructure should both implement the lessons from past disasters and account for future climate-worsened calamities. The choices we make now—especially with regard to infrastructure and land use—can help us thrive even as climate change unleashes unfamiliar weather extremes on the planet.

Seller beware

Any law student can tell you that *caveat emptor* is Latin for "let the buyer beware." For centuries, courts followed the principle of *caveat emptor* when buyers complained that goods they had purchased had defects. But over time, as commerce grew more sophisticated and complex, and after the unregulated sale of securities contributed to the 1929 stock market crash, disclosure of risk emerged as an important tool for ensuring integrity in markets. The paradigm shifted to *caveat venditor*, "seller beware," with new expectations for publicly traded companies to disclose "material information," meaning information that would likely affect the perceived value of stock. In the midst of the Great Depression, the U.S. Congress had created the Securities and Exchange Commission (SEC) and passed the Truth in Securities Act of 1933, measures designed to put teeth in the requirement that publicly traded companies disclose material information.

By the time I became special assistant to President Obama at the White House, it was clear to me that the risks posed by accelerating climate impacts had become material to the bottom lines of many companies. After all, new extremes had begun to upend operations by disrupting supply chains and physically damaging facilities, while the transition to clean energy threatened financial loss for companies heavily invested in fossil fuels and related industries. I knew from my years prosecuting criminal securities fraud that most publicly traded corporations want to steer clear of having to defend to SEC investigators decisions not to reveal information. In 2015, I sought out Brian Deese, who then served as the president's counselor and headed the White House climate team. He expressed enthusiasm for the idea of beefing up SEC disclosure guidance to clarify that climate risk constituted material information. Several months later, word came back to me that the SEC had declined to adjust their existing guidance to reflect growing climate risk. The SEC's inertia reminded me once again that the exponential growth of the climate threat was not yet fully appreciated.

Outside of government, meanwhile, a chorus of finance groups have urged robust disclosure of climate risk. Those groups include the CDP (formerly, the Climate Disclosure Project), the Climate Disclosure Standards Board, the Global Reporting Initiative, the International Integrated Reporting Council, the Sustainability Accounting Standards Board, and the Task Force on Climate-related Financial Disclosures (TCFD), which has become the dominant reporting mechanism for climate risk. In 2017, a group of central banks got involved as well, setting up the Network for Greening the Financial System to urge governments to achieve "robust and internationally consistent climate and environment-related disclosure."[3]

Yet most businesses still misdiagnose climate risk and lack comprehensive strategies for resilience, according to a report prepared for the Global Commission on Adaptation in 2019.[4]

A survey of over 1,100 large companies found that only 4 percent disclosed information in line with at least ten of the 11 TCFD recommendations. A study by the International Monetary Fund concluded that 2019 equity valuations across countries failed to reflect any of the projected changes in natural hazards contained in the commonly discussed climate change scenarios.[5] Meanwhile, the world's top 33 banks allocated close to three-quarters of a trillion dollars to fossil fuel financing that same year.[6] In the United States, researchers concluded that markets had failed to incorporate risks from climate-related flooding. If they had, there likely would have been less development in floodplains.[7]

As one data analytics firm gently put it in 2020, "getting to grips with the finer details remains a significant challenge."[8] Voluntary disclosure has led to inconsistent, idiosyncratic, and unreliable results. Disclosure of climate risk, it seems, is still in its infancy. In light of all these problems, how do we move forward? Regulation.

Given the desultory results from voluntary efforts, a number of countries have, on their own initiative, mandated disclosure. France kicked off the effort in advance of hosting the 21st UN Conference of the Parties that brought the world the Paris climate accord. The French Parliament passed legislation requiring large pension funds and insurance companies to disclose their climate risks. New Zealand took the lead in 2020, becoming the first nation to introduce a mandatory climate-related risk disclosure framework across its entire financial system. "What gets measured, gets managed," said James Shaw, the country's minister for climate change, in explaining why mandatory disclosure was warranted.[9] Great Britain quickly followed suit, announcing it would require financial firms and large companies to disclose risks from climate change by 2025. Canada used its pandemic stimulus to further the cause by conditioning large businesses' receipt of pandemic recovery funds on climate disclosure. In 2021, the European Union announced a suite of measures, including creation of a common "climate taxonomy" and establishment of new reporting requirements that apply to nearly 50,000 EU companies. That same year, the U.S. government signaled it too would consider financial regulation to address mounting climate risk. But for climate disclosure to work, the standards must compare apples to apples. The proliferation of schemes on a national level, rather than across the board internationally, will sow confusion, making it nigh impossible for investors to make meaningful comparisons of the information disclosed. With international standards guiding the process, however, disclosure would yield the intended result—informing investors of the risks posed by global warming and, thus, driving more informed decision-making about where to put their money. Global standards could, in the opinion of Michael Bloomberg, former New York City mayor and founder of Bloomberg News and of the TCFD, "prove to be one of the most important turning points in the global fight against climate change."[10]

Better crystal balls

Whether disclosure improves or not, one approach that could help deepen the understanding of climate risk within companies and governments is scenario analysis. The basic idea, pioneered for military planning by Hermann Kahn, a researcher at the RAND Corporation headquartered in Santa Monica, California, is to use a set of defined stories to examine possible future outcomes. Kahn borrowed the name "scenario" from the Hollywood film studios located just a few miles away from RAND's beach-front offices. "Scenarios," he wrote in the 1960s, "are one way to force oneself and others to plunge into the unfamiliar and rapidly changing world of the present and of the future."[11] Because future events may not resemble what has occurred in the past, scenarios can "stimulat[e] and disciplin[e] the imagination."[12]

In the more than half century since Kahn hatched the concept, the use of scenarios has migrated from military planning to a whole range of industries, including fossil fuel production. The Shell Co., for instance, was an early convert, using scenarios to inform their business strategy since the 1970s. In 1998, Shell researchers created a climate change scenario centered on a huge storm pounding the U.S. Atlantic coast so badly that it sparked young people into climate activism, the government into regulating fossil fuels, and plaintiffs seeking judicial relief into courthouses. The researchers named the scenario "TINA," an acronym for "There Is No Alternative" to adapting. Their assumption was that "only a crisis can lead to change," as they put it in an internal 1998 memo uncovered by the Dutch news organization *De Correspondent*.[13]

The use of scenarios has dominated climate scientists' analysis of how different levels of emissions over time could affect projected climate impacts in the UN Intergovernmental Panel on Climate Change (IPCC) reports. The TCFD has also advocated for the use of scenarios in the private sector to evaluate climate risk. Scenarios have even found a foothold in pandemic planning. In 2006, the *Harvard Business Review* published a special report entitled "Preparing for a Pandemic." The report recognized that a pandemic differs fundamentally from other more traditional business continuity risks. It's not a discrete event, but rather "an unfolding global event." The authors recommended scenario planning for gauging potential effects on product demand and company financials.[14] When the COVID-19 pandemic did in fact arrive, it drove "a renaissance" in the use of scenarios, with companies and governments seeking to imagine what the future could bring.[15]

Likewise, worsening climate extremes have sparked interest in scenario-based planning (Figure 5.1). After the unprecedented Black Summer bushfires in 2019–2020 in Australia, the state of New South Wales conducted an investigation to determine better ways to reduce fire risk. The investigators recommended greater use of scenarios. Scenario planning, they explained, would allow the government to "consider the potential for more significant, more extreme events (including worst-case scenarios), to quantify the risk, to understand the preparation required to respond . . . and

FIGURE 5.1. Scenario planning diagram for the National Park Service shows future threats to forests from droughts, wildfires, flooding, extreme precipitation, and coastal erosion.

Source: National Park Service, "Climate Change Scenario Showcase."

to identify the policy or other interventions that could exacerbate or reduce that risk."[16]

In the decades since RAND pioneered the scenario approach to decision-making, their scientists have sought to make the approach even more powerful in the face of what they call "deep uncertainty." Rob Lempert has led that effort, developing a concept named Robust Decision Making, or RDM. I met Lempert years ago over bowls of steamed mussels in a strip mall on Wilshire Boulevard in Los Angeles. He was dating a childhood friend of mine, whom he later married, and we have been friends ever since. Lempert is a physicist by training, but politics runs in his blood—his mother, brother, and sister have all held elected office.

In 1992, shortly after the IPCC issued its first reports, he and a few of his colleagues won an internal RAND competition for a research project. He decided to pivot his work to climate adaptation, a field that involved deep uncertainty as to how, where, and when climate impacts would unfold. He realized that humans could use computers to test strategies against thousands or even millions of plausible paths into the future. Because RDM uses multiple scenarios, it can identify robust adaptation strategies that will do "well enough" across a wide range of plausible futures. Decision makers have turned to RDM repeatedly to better manage all the unknowns about climate change that accompany policymaking. For example, the U.S. government has used RDM since 2012 to assess water allocations under changing climate conditions along the Colorado River, a river system that provides water to 30 million people.

Neither scenario planning nor its cousin RDM provide an infallible crystal ball, however. The value of these types of tools, according to Margaret McCown, a top war game designer for the

U.S. military, comes from "creating that space for people to think through problems in a hypothetical set of circumstances."[17]

First steps first: Screen for resilience

I trace my interest in climate change to one of my first assignments at the Department of Homeland Security (DHS): creating a climate adaptation plan for the department. And I trace my appreciation for the mechanics of building climate resilience to one of my early assignments at the White House: ensuring that overseas development efforts consider climate risk.

For the latter, I need to thank Lisa Monaco, who served as President Obama's homeland security advisor on the National Security Council (NSC). In mid-2013, she offered me a position on the NSC to assist her with her climate change responsibilities, in particular her role as cochair of a newly created interagency council on climate preparedness and resilience. The council first convened in late 2013 in the Old Executive Office Building's Indian Treaty Room, a marble- and tile-laden extravaganza used by the White House for ceremonies and significant meetings.

As the meeting wound to a close, a hand shot up. Loren Labovitch had a question. Labovitch worked for the Millennium Challenge Corporation, a federal international development agency created by President George W. Bush to reduce poverty through economic growth. Labovitch said that he was worried about the effects climate change would have on international development efforts. He feared that the failure to consider climate risk in development programs would lead to both wasted resources and wasted opportunities. Without missing a beat, Lisa Monaco replied, "We'll take that." Thus, the assignment became mine.

Working with Leonardo Martinez-Diaz, then a deputy assistant secretary at the Treasury Department and with whom I would later coauthor the book *Building a Resilient Tomorrow*, and Kit Batten, the senior climate advisor at the U.S. Agency for International Development (USAID), we formulated a plan to assemble a working group drawn from international development and science agencies. The task? To make sure that overseas investments would withstand future climate impacts. We wanted to make sure that climate-worsened flooding, for example, wouldn't wash away a newly constructed road or that a change in the geographic spread of mosquitos wouldn't render an antimalarial program ineffective.

Once assembled, the group quickly got down to business, a fact that surprised me at the time. Having spent decades in both the federal and California state governments developing policy, I am reminded of the lesson Upton Sinclair taught the world. You don't want to know how sausage or, in many instances, policy is made. All too often, there are too many compromises, egos, special interests, and delays that prevent legislation or an executive order from crossing the finish line. But this time, the policymaking process felt different.

Perhaps it was because the area of climate resilience was so new or that this particular project would involve so little money, but, for whatever reason, the group remained focused on the task at hand and quickly hammered out a proposed approach. Going forward, U.S. agencies providing overseas assistance would screen their investments to determine if they were vulnerable to climate-worsened events and, if they were, take appropriate action to reduce the risk. At the end of 2014, President Obama signed the executive order bearing the government-worthy title of "Climate-Resilient International Development."

Recognizing that climate impacts "threaten to roll back decades of progress in reducing poverty and improving economic growth in vulnerable countries, compromise the effectiveness and resilience of U.S. development assistance, degrade security, and risk intranational and international conflict over resources," the order requires the integration of climate resilience considerations into all U.S. international development work and, more specifically, the screening of investments and strategies for climate risk.[18] By the end of 2016, USAID routinely required all projects and activities to consider climate vulnerability.

Screening as a first step toward climate resilience is not a unique approach. Private sector experts recommend doing the same. For example, the analytic firm Verisk Mapelcroft recognizes the value of high-level screening of climate risk before diving into more complex scenario analysis. The World Bank also believes screening projects for climate change risk helps safeguard the long-term success of its development efforts. Other financial institutions, including the European Investment Bank and the Inter-American Development Bank, have also implemented screening requirements in their processes. Nowhere is it more important to engage in screening for climate risk than when considering long-lived infrastructure projects.

Getting infrastructure right

Investing in resilience can save lots of money in the long run. The COVID-19 pandemic has come at an almost unfathomable cost—both in lives lost and in economic losses, which as of 2020 were estimated between \$9 and \$33 trillion. If the world prepared, it could prevent, or at least greatly reduce, this staggering level of loss. According to an analysis by the international consulting firm McKinsey & Company,[19] global investments of \$70 to \$120 billion over two years in such preventive measures as faster response capabilities, better disease detection abilities, and larger critical care capacity could substantially reduce the damage from another pandemic.[20] When it comes to climate change, investments in infrastructure resilience hold similar promise for big payoffs. The World Bank estimates that investments in infrastructure resilience can yield a return of \$4 for every \$1 invested.[21] Going forward, infrastructure has to play a leading role in climate resilience planning.

Infrastructure lies at the center of human civilization. It supplies the foundation that makes society possible, enabling myriad essential services, namely communications, waste removal, water treatment, transportation, and power. Our lives literally depend on it. In the coming decades, nations will have to invest deeply in infrastructure, given the need to repair, upgrade, and replace aging systems like bridges, roads, and sewers to meet future challenges. The Organisation for Economic Cooperation and Development pegs global infrastructure investment needs at \$95 trillion by 2030. That's more than the value of the entire current stock of infrastructure and an average of more than \$6 trillion per year. That's a lot of money, so it's crucial that it be spent with resilience in mind.

If infrastructure cannot withstand coming extremes, it can cause cascading failures that leave entire communities and regions exposed to harm. A 2019 study estimated that in Vietnam, one of the world's fastest-growing economies, climate-related disruptions to critical road networks could cause damages of close to $2 million per day, while railway failures could lead to losses as high as $2.6 million per day.[22] Worldwide, infrastructure disruptions from natural hazards, poor maintenance, and mismanagement already cost businesses close to $400 billion and families $90 billion in low- and middle-income countries every year, according to the World Bank.[23] Those costs will mount as climate change brings more severe impacts causing more disruptions to critical infrastructure services.

One of the challenges of building resilient infrastructure is planning for climate extremes that may not be experienced for decades. Infrastructure typically has a long service life. Take hydroelectric dams. They are built to last 70 to 100 years. Likewise, the expected service life of a bridge is 50 to 100 years. It's even better if infrastructure lasts beyond its expected service life, as Commanding General Thomas Bostick of the U.S. Army Corps of Engineers once told me. The Romans managed to construct bridges still in use today, totaling over 900 in 26 countries.

The expectation is that infrastructure built today will have to stand up to new extremes that could occur long into the future. Yet traditional cost-benefit analysis does not adequately value the potential benefits that would come from reducing the damages caused by those future climate extremes. David Espinoza worries about this problem and how it could leave communities unprepared.

Espinoza has spent over a quarter century as an engineer designing large infrastructure projects around the globe. Wiry and fit, he can easily be pictured scrambling over boulders to examine some remote building site. He worries that companies do not yet invest sufficiently in resilience. He has witnessed time and time again resilience features cut from plans because the cost-benefit analysis does not value resilience provided decades in the future, or because the owners and operators believe that the added costs outweigh the benefits to a risk they think may not materialize. Espinoza knows that communities will end up investing in resilience eventually. But, "once the concrete is poured," he says, "it becomes more expensive to add." To avoid the risk of critical infrastructure failing prematurely or having to perform expensive retrofits, governments should provide fiscal incentives to push private investors to take a longer view and adopt cost-benefit analysis methods that quantify gains from avoided emissions and future resilience.

How much does building in resilience cost? It depends. Installing a larger drainage ditch to handle higher flows from heavier rainfall would come at a relatively modest cost and yield benefits of up to $200 in averted flood damages for every $1 spent. Painting roofs white to reflect heat and thus keep buildings cooler is also an inexpensive option. Nature-based resilience solutions—such as increasing tree canopy to provide cooling or restoring wetlands to buffer storm surge—likewise cost comparatively little. But other approaches carry big price tags. Elevating assets, like raising a railway track to avoid flooding, could balloon costs by 50 percent, according to the World Bank.[24] Achieving higher levels of resilience can also drive up the expense significantly. A study after Hurricanes Irma and Maria swept across Puerto Rico revealed that upgrading electrical transmission systems to withstand a Category 3 hurricane would hike costs by 3 to 40 percent; upgrades for a Category 4 storm would drive the costs up by 24 to 70 percent.[25]

Given these kinds of additional costs, communities may need to make trade-offs in the level of resilience they seek. To contain expenditures, they could consider adding resilience in incremental steps. For example, Japan Railways, concerned about derailments caused by rails buckling in severe heat, bumped up the heat performance standards for its railroads by just 5 degrees Celsius (from 60 to 65 degrees Celsius) for future acquisitions rather than taking a bigger leap.[26] Doing so is consistent with an approach advocated by adaptation experts. An adaptation pathway is a strategy for making decisions among various adaptation choices in the face of the many uncertainties posed by climate change. The process encourages decision makers to follow a defined sequence of steps over time, with predetermined triggers governing when the next step is required. The goal is to keep evolving so that plans can adjust as knowledge about climate change accumulates and circumstances change.[27]

Some of the most famous long-term adaptation programs have followed the adaptation pathways approach. Consider the Thames Barrier, which protects the city of London from flooding (Figure 5.2). The project took three decades to build, but it is far from finished. Given the enormous cost involved, the planners developed a decision pathway that takes the project up to the year 2100, and which includes triggers for decision points as to when to alter the structure based on sea-level-rise measurements. The pathway also includes "low regret" options that do not rule out constructing an entirely new barrier, after planners have exhausted other more cost-effective measures and when sea-level rise reaches a certain level.

Using approaches like adaptive pathways could save nations and communities trillions of dollars as they make infrastructure investments in the coming years. It also buys time. But it's not just

FIGURE 5.2. The Thames River Barrier protects London from flooding by opening and closing based on tidal flows.
Source: AC Manley / Shutterstock.com.

how we build infrastructure that will matter going forward; it's where we decide to build too.

The choices we make today matter

There is probably no country in the world that has devoted more time and resources to the study of flooding than the Netherlands. Over 60 percent of the country is vulnerable to flooding, from the sea as well as rivers that run through the country. The Dutch historically built dikes and other infrastructure to keep out the water. But in 1993 and 1995, rivers burst their banks, forcing the evacuation of 250,000 people and one million livestock. That flooding drove the Dutch to rethink their approach to living with water.

They realized that building ever-higher dikes would not solve the problem, particularly with climate change. The Netherlands needed a better plan for flood resilience.

In 2006, the Dutch government began its flagship "Room for the River" program, a $2.5 billion project over more than ten years. Rather than reinforcing or replacing existing dikes to manage the waters of nearby rivers, the program gave the rivers more room by deepening them, lowering floodplains, and removing polders, which are tracts of land surrounded by dikes. Expanding the floodplains provided additional lands to act as "natural water sponges," while allowing for the creation of more parks and recreation areas. It simultaneously reduced the flood risk to adjacent communities. Some families had to move to other locations to free up the necessary land, but the bottom-line result is that the Room for the River program improved flood safety for nearly four million people.

The program's success rested at least in part on its inclusivity. The government worked with stakeholders at all levels— municipalities, provinces, regional water authorities, and local residents and businesses. The national government set the overall parameters of the effort, but local and regional decision makers determined the specifics. The extended process engaged different agencies working on different issues, ranging from water safety to agriculture at local, regional, and national levels. The agencies in turn sought input on how to improve communities and better support those families forced to relocate.

Even before completion of the Room for the River project, the Dutch moved onto the next step in their resilience planning with a nationwide initiative called the Delta Programme. The government's stated goal for the project was to "prevent a repetition of the 1953 flood disaster and of river flooding" in the 1990s, two calamities that live on in Dutch memory.[28] The program was designed to enhance flood resilience, ensure sufficient freshwater supplies, and climate-proof the country's planning for land use. It also had the explicit goal of serving as a model of good practice for international water management. And that's one of the reasons that the Dutch remain global leaders on climate resilience. They look ahead to find better ways to build resilience to changing circumstances.

According to the global reinsurance company Munich Re, flooding has cost the world more than $1 trillion in losses since 1980, and the risks are increasing from both climate change and continued development in at risk areas. Better land-use choices would reduce those risks. That could mean intentionally opening up more areas to flooding as the Dutch did. It could mean restricting development in areas prone to flooding like Ontario did after Hurricane Hazel. The United States, in contrast, has taken a mostly hands-off approach, leaving floodplain management largely to local communities. As a consequence, census data shows the number of people living in floodplains in the United States has exploded, increasing by 14 percent from 2000 to 2016, at a rate faster than areas outside flood zones. This explosion in development has occurred even as the risk of flooding from sea-level rise, storm surges, and heavy rainfall has worsened due to climate change.

The United States is far from alone in its predilection for building in floodplains. One in ten new homes built in England since 2013 is in a flood-prone area. As the British government pushed to alleviate the country's housing shortage, the number of new houses built on land at the greatest risk of flooding grew from

9,500 homes in 2013 to 20,000 in 2017–2018. In Calderdale, a West Yorkshire valley hit by multiple floods in recent years, one in five new homes was built in the 100-year floodplain, which has a 1 percent or higher risk of flooding in any given year.

It's not just flooding that requires adjustment of land-use policies. Other climate-fueled hazards come into play as well. Consider wildfires. As many nations around the world have learned recently, the lands many live on have become more flammable. Australia, Canada, Chile, Greece, Portugal, Russia, and the United States have all suffered record destruction from wildfires since 2000.

Clearly the nature of fire is changing. For example, in the western United States, the number of fires has mushroomed. Over 60 percent of wildfires between 1950 and 2017 have occurred after 2000, with 2020 reigning as the most active fire year in recorded history. California wildfires in 2020 broke "almost every record there is to break," according to Daniel Swain, a climate scientist at the University of California, Los Angeles.[29] Before 1970, a megafire, a fire that burns more than 100,000 acres or 156 square miles (404 square kilometers), simply wasn't part of the firefighting vocabulary. In 2020, California suffered its first recorded gigafire, a blaze that burned over a million acres.

In Australia, the size and nature of fires have also changed, rendering firefighting techniques and strategies that have worked well in the past less effective. During the country's Black Summer of 2019–2020, fires burned in forested regions on a scale never before experienced in its recorded history. Firestorms forced some people to evacuate to nearby beaches to escape the flames. The fires scorched an area larger than Cambodia, destroyed close to 6,000 buildings, and killed over 30 people and up to three billion animals. The conflagrations took "even experienced firefighters by surprise."[30]

Sadly, areas that have burned already face the prospect of fires returning in the near future. Almost a third of the land burned from 1950 to 2017 in the western United States was "burn on burn." Close to one-fifth of the 2020 fires burned in areas that had also burned in the previous two decades. As firefighters will tell you, if it has burned before, it risks burning again. Similarly, Australian authorities predict that any reductions in future fire risk as a result of the 2019–2020 fire season are "partial and temporary," with the risk increasing within the next six years to levels equal to or exceeding those existing before the Black Summer.[31]

Wildfire, like flooding, renders some places that were previously relatively safe to live more dangerous. Australia's 2009 "Black Saturday" fires, which torched thousands of homes and over a million acres, led the state of Victoria to fast-track new building requirements for fire-prone areas. But a decade later those measures proved inadequate in the face of the intensity of the Black Summer fires of 2019–2020. The bushfire code will undergo a new assessment.

The state of California has also struggled with finding ways to make sure people can live safely in wildfire-prone areas. It adopted the most stringent wildfire code in the United States in 2008. Yet when California suffered from record-breaking fires in 2017 and 2018, many homes built to that code were incinerated.

The losses in these wildfires highlight why decisions about where and how we build matter—just as they do when it comes to flooding and other climate-worsened risks. But communities, even in some of the richest countries in the world like the United States and Australia, still often make those decisions without adequate consideration or understanding of the mounting risks. In both nations, many older homes do not meet current minimum standards for fire protection, and there is no requirement to upgrade them or relocate residents from fire-prone areas. Piling onto the problem is the fact that a homeowner complying with existing building standards in an at-risk area can lead, according to a recent Australian review of the Black Summer fires, to a "false sense of security [that] it will be safe to live there."[32]

Adjusting land-use practices to account for climate risk would reduce future damage. Restricting land use, however, remains politically fraught. In the decade from 2009 to 2019, California built more than 10,000 homes in areas where the risk of wildfire is high.[33] But as of this writing, the state has rejected the idea of limiting development in high-risk areas, something its governor claimed would conflict with California's "wild and pioneering spirit."[34] Similarly, in Australia, in one of the states hardest hit by the Black Summer fires, there is no legal prohibition against "development in areas where bushfire risks are too great and cannot be mitigated."[35]

To respond to the growing risks, building codes should undergo an overhaul as well. They need to help preserve the structure's ability to function post-disaster rather than just protecting the life and safety of its occupants. In other words, codes should cover a building's performance during and after calamity. Ensuring that buildings and infrastructure can function as intended in the wake of a disaster assists communities and households in recovering more rapidly while saving lives, livelihoods, time, and money. Performance building codes allow for greater innovation in design and material, as opposed to prescriptive codes, which require strict adherence to minimum construction standards, such as the size of nails and the distance between studs in wood-frame buildings. Performance-based standards can take into account site-specific conditions. They can lead to more aesthetic, functional, and economic solutions to focus on the ends rather than the means to determine what a structure is required to do, instead of prescribing how it should be built.

Japanese engineers have pioneered performance codes for earthquakes to address their country's high seismic vulnerability. The codes are designed to ensure that buildings can not only withstand earthquakes but also be used immediately afterward. And by addressing how a building should perform, rather than prescribing how it is built, the codes give designers and builders more leeway in selecting construction methods. Performance building standards have proven to be very effective.

I vividly remember the story a Japanese acquaintance shared with me about the 2011 Tohoku earthquake described in Chapter 1. He and his family had gathered the night before the quake at his apartment in a high-rise building in the city of Sendai. They had celebrated by drinking a bottle of champagne. The next day, they left on a trip, leaving the empty champagne flutes standing sentry on his kitchen counter. Not long after, the earthquake struck the city of Sendai. The epicenter was less than 50 miles (80 kilometers) away from his apartment. When his family returned, the apartment was unscathed, and the tall, slender champagne flutes were still upright.

Other earthquake-prone countries, including Peru, Turkey, Chile, China, Mexico, and Italy, have followed Japan's performance-based approach to building codes to varying degrees. New Zealand, an early adopter of performance codes, explicitly requires that essential structural elements like roofs, walls, and floors be built to last at least 50 years. That means that designers and builders must account for altered conditions brought by climate change over the course of the building's life.

As climate extremes batter buildings with greater frequency, nations need to develop performance-based standards that both save lives and ensure the continued function of structures post-disaster. If communities build homes only to fail in the next, even worse storm or wildfire, they will struggle to recover economically when people are forced to relocate to other areas in search of housing and functioning businesses. Performance-based building practices will help communities bounce back faster just as they did for Sendai and my acquaintance's family in his high-rise apartment. Of course, the added resilience must be weighed against any added cost of construction. But finding ways for construction to better account for mounting risk should remain a goal going forward; focusing on building performance could help achieve that.

The reality is that land-use as well as construction standards matter deeply when it comes to protecting against future damage. If people choose to live in risky areas, national governments should not routinely subsidize those decisions or routinely risk the lives of emergency personnel when disaster strikes. Rethinking land-use and construction practices remains core to jumpstarting climate resilience. So does sharing the lessons learned. As countries and communities adopt new design options and land-use policies, sharing those across borders would allow for speedier spread and adoption of emerging resilience practices. John Roome, senior director for climate change at the World Bank, put it succinctly: "It is not about spending more, but about spending better."[36]

Commemorating the loss

One way to drive greater preparedness is to encode past disasters in the DNA of everyone. Disasters happen. Disasters cause loss of life, despite the best of preparations. But disasters can help spur even better preparedness going forward. And helping people remember disasters can galvanize nations to focus on being better prepared. At least that is what Ron Thieman, director of Deltares, an independent water institute in the Netherlands, believes.

Thieman, a tall imposing presence with a booming voice, speaks with a barely contained excitement about how Dutch researchers continue to learn the best ways to live with water. As we drank cups of espresso and nibbled butter cookies in The Hague's Peace Palace, built with the money of American railroad magnate Andrew Carnegie, and home to the International Court of Justice, Thieman leaned in to explain how every Dutch citizen knows about the devastating storm of 1953. The storm spread floodwaters across wide swaths of the country and killed 1,800 people. He told me that schoolchildren learn about it and that, every February 1, the country commemorates the *Waternoodramp* or the "water emergency disaster." He excitedly added, "We even have a museum dedicated to it. You must see it!"

Japan, a country also known for its preparedness, has likewise invested in remembering its "disaster heritage." As we saw in Chapter 1, for centuries the Japanese placed stones to mark the height of tsunami flooding as a warning to future generations. They have also created memorials to more recent disasters. In September 1959, the biggest typhoon Japan had ever experienced in its recorded history made landfall, killing 5,000 people. After that, the Japanese government determined that, going forward, it would find ways to increase public awareness of the need to prepare, including through the commemoration of past disasters.

As a first step, the government turned to what was then the not-so-distant past, the magnitude 7.9 earthquake that struck Tokyo on September 1, 1923. The shaking caused a fire to ignite that quickly consumed tens of thousands of the city's wooden buildings. Over 100,000 people lost their lives. Going forward, Japanese authorities declared, the nation would recognize September 1 as Disaster Prevention Day. The goal? To reduce the death toll in future disasters. Then, almost four decades later, another disaster pushed the nation toward even greater preparedness.

A magnitude 7.2 earthquake struck the southern city of Kobe in the early morning of January 17, 1995, killing over 6,000 people, making about 300,000 residents homeless, igniting hundreds of fires that incinerated the equivalent of 70 city blocks, and causing nearly $200 billion in economic losses. City government offices suffered damage, leaving residents adrift. Some blamed the government's poor disaster preparedness for the widespread destruction. In response to the devastation, over a million people converged on Kobe to aid the recovery.

According to one volunteer, the outpouring of help from citizens gave "rise to a quiet revolution in the country" of increased volunteerism.[37] Eleven months after the earthquake, the Japanese government declared January 17 as national Disaster Prevention and Volunteerism Day to encourage future voluntary disaster preparedness and relief efforts. Like the Dutch government, the city of Kobe also created a museum, as well as a park, to educate the public. And following the massive earthquake and subsequent tsunami in 2011, November 5 became, not surprisingly, Tsunami Readiness Day. The city of Sendai, recognizing that the number of residents who had experienced the 2011 Great East Japan Earthquake was already decreasing, preserved the ruins of an elementary school destroyed by the tsunami to make plain the threat of tsunamis to future generations.

These commemorative efforts keep alive the discussion of "no more" moments and a reminder of the need to prepare actively circulating in the media, in the schools, and at the dinner table. They make it easier for "companies, neighborhoods, or households to address each of these topics again and again over the years."[38] Should nations fail to remember, they may find themselves ill-prepared in the future.

Consider the 1918 Spanish flu pandemic. The world suffered "mass amnesia" about the event.[39] Barely any memorials exist, and those that do were created decades after the event. Little was written about it. The first significant account did not appear until 1976. Then in 2004, author John M. Barry published a book telling the story. In noting the paucity of writing about the pandemic, he observed that "people forget the horrors that nature inflicts on people, the horrors that make humans less significant."[40]

As communities suffer tragedy, finding ways to remember past tragedies may help them better imagine what could occur in the future, just as Ron Thieman assured me it had for the Dutch. Recalling disasters combats availability bias, or the tendency of humans to give more weight to recent events when making decisions, thinking that those events represent the probability of a future event occurring. Availability bias can cause people to forgo buying insurance coverage in the absence of a recent disaster, for example. But after a disaster strikes, they tend to purchase insurance assuming that the recent event reflects the probability of a future event. As the disaster fades from memory, however, so does the interest in paying for more insurance.

Availability bias applies to corporations too. In 2018, the insurance broker Marsh and the reinsurer Munich Re created a parametric insurance policy to cover pandemic-related losses. But not a single business bought it. Without the memory of a recent global pandemic to incentivize preparedness, businesses declined the coverage. Once the COVID-19 pandemic hit, causing mass disruptions in the global economy, demand for the pandemic insurance surged as companies searched for relief from coronavirus-related losses. However, it was too late. As the senior vice president of Marsh's Alternative Risk practice put it, if companies had bought the coverage in November 2019—before the coronavirus emerged—losses would have been covered, "but you can't buy insurance for your house when it's already on fire."[41]

By commemorating a past disaster, communities and governments can help keep the prospect of calamity—and, most important, the need to prepare—foremost in people's minds. Sharing stories of past calamities reminds people of the risk and critical need to continue to build resilience. After all, that is what two of the world's most disaster-ready countries—the Netherlands and Japan—have chosen to do. Public commemoration can help spur readiness in other ways too. Once nations claw their way out of the COVID-19 pandemic, they should establish national days dedicated to public health preparedness. They can follow the example of Jersey City, New Jersey, which in late 2020 committed to creating a COVID-19 monument by constructing a memorial wall listing the names of all residents killed by the coronavirus and planting more than 500 trees in their honor.

Taking time to appreciate prior losses will help build greater resilience to the next big disaster. As one Japanese official stated at a 2017 UN disaster conference, "If we deepen the understanding of the next generation about disaster risks, it will help to decrease the impact of disasters. Memories can be lost so they must be passed on."[42]

CHAPTER 6

MARRY MITIGATION AND ADAPTATION

> Millions don't rally to the banner of uncertainty.
>
> —*George Packer, American novelist*

The COVID-19 pandemic is a story of missed opportunities. Determining where and how the world stumbled so badly could occupy a whole generation of think-tank scholars drafting white papers, university professors hosting symposia, and graduate students huddled over dissertation manuscripts. Answers to questions such as "How did China's 17-day silence before news of the novel coronavirus leaked affect the disease's spread?," "Why did governments largely ignore health experts' warning that for pandemics 'it's a question of when, not if'?," and "Can reedy thin supply chains work in a time of disaster?" could lead to greater preparedness for the next pandemic.

What we do know already is that the spread of COVID-19 led to the broadest economic collapse globally since 1870. The pandemic threw tens of millions of people into poverty. And as the disease forced political leaders to open national purses ever wider, pouring out cash to keep their economies from crumbling, their initial response shoved action on an even greater long-term threat—climate change—deeper into the background. The pandemic was a test of governments' and businesses' commitments on climate change, argued the executive director of the International Energy Agency (IEA). Judging by their actions, it was a test that, as of this writing, they mostly flunked or decided to retake on another day.

By June 2020, Bloomberg News estimated that 50 of the world's largest economies had plans to commit close to $12 trillion to stimulus measures to try to limit the economic pain from the pandemic. But less than 0.2 percent of the funding was aimed at climate-related and environmentally friendly initiatives such as renewable energy.[1] An analysis by the economic consulting firm Vivid Economics found that most of the G20 countries, a group of the world's leading economies, had "largely failed" to use the crisis to advance green measures. Even worse, G20 nations had actually provided unconditional support to polluting industries and rolled back environmental measures.[2]

Finding answers for why most nations missed the opportunity for a green recovery, like finding answers to why the pandemic got so out of control, is an important task. Examining the factors that drove political leaders to ignore the climate threat, even as it loomed over their shoulders, could expose barriers to future progress on climate change, barriers that must be overcome. Did concerns about costs drive governments to embrace brown rather than green? Did perceived uncertainty about climate risk deter action? Did opposition from entrenched industrial interests quell political support? My own view is that each of these factors played a role. But I also see another reason that progress on climate change lags: the long-standing division within the climate community itself.

By the time I received the assignment to lead the development of the Department of Homeland Security's (DHS) first-ever climate adaptation plan in 2009, stacks and stacks of reports warning of the dangers posed by climate change were already piled high in bureaucrats' offices around the world. The UN Intergovernmental Panel on Climate Change (IPCC) began issuing reports in 1990 and never stopped. In 2000, the U.S. government released its first national climate assessment. In 2009, it was in the process of scoping its third. Climate change had finally moved out of its eddy as a niche policy topic for the distant future into the mainstream.

The DHS assignment gave me, along with other members of our task force, the chance to learn from these reports and to hear from climate scientists, military strategists, and infrastructure experts regarding climate risks. As we examined the evidence, my training as a judge told me that more than enough proof existed to take action to address the threats under any legal standard. Cutting emissions had to occur to avoid the very worst impacts of climate change. But with impacts already evident, adaptation had an important policy role to play as well. To tackle climate change, policymakers and businesses should address both sides of the climate change equation. Or as President Obama's climate advisor John Holdren used to remind me, "We must avoid the unmanageable and manage the unavoidable."

Few decision makers in the public or private sectors, however, have responsibility for both adaptation and mitigation efforts. Reflecting this division, attention to climate issues remains sprinkled across academic disciplines and government agencies. At the national level, foreign, environmental, and energy ministries handle mitigation issues, while agriculture, forestry, and infrastructure ministries address adaptation. In 2017–2018, only 2 percent of climate financing went to projects with dual mitigation and adaptation objectives.

The separation of those who work on adaptation and those who work on mitigation hampers effective action to tackle climate change.

There is no natural mechanism to unlock the flow of information from one group to the other, so communication suffers. As a result, according to the IPCC, different communities of interest have formed, each pursuing unrelated research and policy discussions. The division can cause a government's or business' focus to scatter among experts who only specialize in one side or the other of the climate equation, thus erecting yet another barrier to productive policies.

Even more worrisome, the division between adaptation and mitigation camps can result in climate policies that actually cause harm, because efforts in one area may undermine efforts in the other. The failure to consider vulnerability to climate-fueled extremes in emissions-cutting efforts can result in "malmitigation," while the failure to consider emissions caused by adaptation projects can result in "maladaptation."[3]

Going forward, decision makers should assess emissions reduction and preparedness choices in view of their overall contributions to addressing climate risk. An easy way to marry mitigation and adaptation is through nature-based solutions, which reduce planetary warming by absorbing carbon while simultaneously protecting against the accelerating impacts of new climate extremes. Harder is the question: who will pay for what needs to get done? Fortunately here too, promising answers have begun to emerge.

Solving one problem while creating another, or "maladaptation"

When it comes to preparing for climate change impacts, many government and community leaders tend to favor engineered solutions involving concrete and steel over nature-based solutions.[4] Their first instinct is to build a wall to protect against rising sea levels as opposed to restoring natural features, such as wetlands or mangrove forests that act as natural buffers. A 2019 review of national legislation in 33 countries revealed that flood-related laws failed not only to consider future risk but also to recognize the benefits of nature-based solutions over "hard engineering measures."[5] Nature-based solutions can provide sinks to store harmful carbon emissions. Producing the concrete and steel used to construct engineered solutions, on the other hand, consumes massive amounts of water and adds more carbon to the atmosphere, which, in turn, leads to further rises in temperature—in other words, "maladaptation." It's time to change the paradigm, reaching for nature first—not last—to address both sides of the climate equation.

While I was in the White House, I observed how quickly people can embrace engineered solutions to ease climate-worsened stressors. In 2014, Obama administration leaders gathered in a conference room in the Old Executive Office Building to discuss ways to combat the drought that then gripped the American West and the dwindling availability of fresh water in some parts of the country. Specifically, the administration wanted to demonstrate to the American public its commitment to improving drought response.

Not long after the policymakers had settled into their seats around the conference table, a voice asked, "What about desal?" "Desal" refers to desalination, a process that can ease water woes by transforming salt or brackish water into potable water. Desal would allow communities to tap into the limitless supply of ocean water along the coastline of the western United States. It would alleviate the water stress from climate-worsened drought. In placing a bigger bet on desalination, the United States would join a group of water-starved countries that had already turned to the process.

Large-scale desalination plants began cropping up in the 1960s. Since then, use of the technology has exploded. Over 18,000 plants in 140 countries now dot the globe, according to the International Desalination Association. Saudi Arabia and Israel have made deep investments in the process. So too has Australia, which went on a desalination plant building spree after the "no more" moment of the Millennium Drought desiccated southern portions of the country in the 2000s. An estimated 300 million people rely on desalinated water for some or all of their daily needs.

Backers of the process press the allure of an inexhaustible flow of fresh water in drought-prone coastal regions. Inland promoters push the use of brackish water to create potable liquid gold. North Africa and the Middle East have experienced an estimated annual 7 to 9 percent growth in desalination installations since 2016. The city of Carlsbad, California, opened the largest plant in the Western Hemisphere in 2015. It treats about 100 million gallons of seawater daily, which provides approximately 10 percent of the water for 3.1 million people. The water from desalination does not necessarily come cheap, however. The Carlsbad plant's water costs twice as much as other sources and impacts the environment negatively in multiple ways.

When desalination was first proposed as a solution at the White House meeting, the discussion did not touch on a factor that affects the other side of the climate equation—greenhouse gas emissions. Existing desalination systems use astounding amounts of energy. It takes about 10,000 tons of oil per year to desalinate

1,000 cubic meters of water (1,000,000 liters or 264,000 gallons) per day. Going forward, choices about desalination, especially in the Middle East, could have profound effects on global emissions.

The IEA estimates that from 2019 to 2040, the Middle East will increase its production of desalinated water 14-fold.[6] The Middle East is among the driest regions in the world with some of the most water-scarce countries. The area already houses 70 percent of the world's desalination plants. In addition, it has vast gaps between renewable water supplies and demand. As of 2015, Saudi Arabia used 943 percent of its available renewable water reserves, while Kuwait used 2,465 percent, and Bahrain used 220 percent. Demand for water will only swell as populations grow and temperatures rise, making alternative means of acquiring water all the more urgent.

To the extent that the plants rely on fossil fuels for energy, desalination as an adaptation strategy contributes to future warming, which, in turn, will add to water scarcity. Consider Saudi Arabia. Flush with fossil fuel energy but little fresh water, the Saudis possess close to a fifth of the world's desalination capacity. Using 300,000 barrels of oil a day, the country devotes 25 percent of its domestic oil and gas production to powering desalination plants. That figure could rise to 50 percent by 2030 according to UN estimates, if the country does not switch to more sustainable energy sources.

Desalination consumes about 0.4 percent of the world's energy supply already, with only 1 percent of desalinated water produced from renewable energy sources. The plants additionally pose environmental challenges, leaving a gallon (3.79 liters) of briny water for every gallon of fresh water produced. If dumped back into the ocean in a concentrated area, that briny water can negatively impact marine life, further imperiling ocean health.

The idea of promoting desalination never picked up steam during the Obama administration. Instead, the White House chose to promote a package of programs to alleviate drought immediately while promoting long-term drought resilience planning, including greater water conservation. President Obama signed an executive order developed by my team at the National Security Council that required federal agencies to assist state and local officials with drought management by sharing data, information, and research; communicating drought risks posed to critical infrastructure; bolstering capacity for drought preparedness and resilience; and supporting efforts to conserve and make efficient use of water. As climate change reduces freshwater availability, however, desal will continue to attract the attention of government and business leaders seated around countless other conference tables around the globe as a possible solution.

When it does, policymakers should recognize the risk that some adaption measures could derail emissions reduction efforts and instead identify approaches that do not create more carbon pollution. Policymakers could, for example, re-examine whether water conservation measures, such as the reuse of wastewater or "toilet to tap" processes, treatment of stormwater runoff, collection and storage of rainwater, restoration of ecosystems to better retain water, and improved irrigation to lessen evaporation, are better solutions for reducing water scarcity than desalination.

If communities still want to go forward with adaptation solutions involving steel and concrete, like building a desal plant, they need to look for ways to use clean energy to power the facility. A 2016 study of a pilot program in Abu Dhabi found that solar-powered desalination was up to 75 percent more energy efficient than the fossil-fueled technology then used in the rest of the United Arab Emirates.[7] Dubai has decided to go big with solar-powered desalination. It has plans to produce over 300 million gallons of potable water per day using solar energy at an estimated savings of $13 billion by 2030. The city of Perth in Australia has turned to wind power for its desalination plant. Whatever choices communities, businesses, and nations make going forward to address climate extremes or to cut emissions, they should examine both sides of the climate equation and continue to search for a win/win.

Going green but forgetting about climate, or "malmitigation"

Neglecting to consider both mitigation and adaptation can lead to unfortunate, and costly, climate surprises when it comes to choices for cutting emissions. Take the energy sector. Natural shocks, such as floods, storms, and drought, already cause 44 percent of power outages in the United States and 37 percent in the European Union, upending economic activity and plunging communities into darkness. New climate conditions—heat waves, water stress, Arctic cold blasts, "rain bombs," and sea-level rise—occurring with greater frequency and severity will test existing energy infrastructure. When energy system planners focused on reducing emissions do not factor in new climate extremes, they can inadvertently create new risks to power generation and even human safety. Consider how the failure to account for climate extremes can curtail nuclear power generation or dim wind and solar energy's effectiveness.

As of 2020, the world had 442 operating nuclear plants. The reactors provide approximately 10 percent of the world's electricity and about 30 percent of all low-carbon power. As nations look for ways to hold the total average global temperature rise to below 1.5 degrees Celsius (2.7 degrees Fahrenheit), international energy organizations—including the International Atomic Energy Agency (IAEA), the American Nuclear Society, the European Nuclear Society, the World Nuclear Association, and the IEA— have touted nuclear power as green. Extending the service life of these older plants could cement a cost-effective opportunity to maintain low-carbon energy capacity, or at least that's what nuclear energy proponents argue.

Just as it is often less expensive to repair than to buy new, extending the service life of a nuclear power plant is way cheaper than constructing a new one. Indeed, in advocating for the long-term extension of existing plants, the IAEA asserts that on a technical level "most existing reactors could be safely operated until they are 60 years or older."[8] In the absence of extensions, it predicts sharp declines in existing nuclear capacity by 2030 in North America and Europe and the retirement of all existing plants by 2060.

But the nuclear plants in the global fleet are old. Two-thirds commenced operation in the 1970s and 1980s, with a then-expected service life of 30 to 40 years. Their builders did not consider worsening climate extremes in their design, construction, maintenance, and operation. As a result, keeping plants operating past their prime carries risks to both energy generation and the plants themselves. When new extremes hit existing plants, they could result in a loss of the critical cooling systems required to prevent nuclear fuel from overheating, which can lead to a meltdown.

France came frightfully close to a nuclear disaster in 1999. Storm surge and strong winds during a high tide had pushed waves from an adjacent estuary over the dikes protecting Le Blayais Nuclear Power Plant. The flooding forced three of the facility's reactors into an emergency shutdown. Fortunately, a central cooling pump remained functional, preventing a nuclear meltdown. The specter of utter catastrophe prompted French authorities to spend 110 million Euros ($122 million) on flood prevention measures like raising seawalls and strengthening dikes as well as creating a program to monitor future climate impacts on power plants.[9]

More than 40 percent of existing nuclear plants are sited near shorelines, making them particularly vulnerable to storm surge, sea-level rise, and intensification of storms. Inland plants face threats from riverine flooding and wildfires fueled by drier, hotter conditions. A fire came dangerously close to the San Onofre Nuclear Generating Station in Southern California in 2014. In addition to these dangers, climate extremes can wreak havoc on the generation of nuclear power. A heat wave in France in 2018 shut down four plants. In another instance, drought shrank the body of water used to cool the reactor, exposing the plant's intake pipes. Too warm waters led to the shutdown of Connecticut's Millstone plant in 2012 for 12 days (Figure 6.1). The largest source of power in the state had to halt operations at the height of summer with air-conditioning demand at its peak. The seawater used to cool the nuclear units had become too warm—something the plant's designers had surely never imagined. The shutdown forced reliance on fossil fuel plants to compensate for the energy shortfall, resulting in more carbon emissions.

As countries seek to eke out more life from nuclear reactors past their prime, the policy debate has focused on the reduction

FIGURE 6.1. Dominion Millstone Nuclear Station, where a reactor was forced to shut down due to increased Long Island Sound water temperatures rendered ineffective for cooling.
Source: JJBers / Flickr, licensed under CC BY 2.0.

of greenhouse gas emissions, nuclear proliferation, and the use of small reactors. It has largely ignored the growing safety and energy security vulnerabilities created by new weather extremes to decades-old plants, designed and constructed with no consideration of global warming. The decision whether to extend a plant's service life rests with individual sovereign nations, yet the consequences of those decisions carry international implications if climate impacts damage nuclear facilities. To date no international entity has assumed responsibility for determining the growing threats reactors face, much less their ability to operate safely in a changed climate. The absence of both oversight and standards, for new and existing plants, with regard to hotter conditions puts international energy security, the environment, and human health at risk.

Other clean energy sources, solar and wind power, likewise face disruption from accelerating climate extremes. Consider power from the sun. Production of solar energy has exploded around the world, promising carbon-free energy for hundreds of millions of people. As the cost of solar photovoltaic (PV) panels, which directly convert sunlight into electricity, has plummeted, nations around the globe have enthusiastically embraced the more affordable technology. Panels have sprouted on homes, commercial buildings, and industrial facilities, as well as in giant "solar farms" that stretch for miles.

The year 2019 set a record for PV deployment. The United States, the European Union, Latin America, the Middle East, and Africa all saw significant gains in capacity. Despite the economic recession caused by the pandemic, the United States experienced almost a 50 percent growth year over year in 2020 in solar power generation. The combination of abundant sunlight and rapidly declining costs for the technology to harness it has turned solar energy into the poster child for a green future. The IEA predicted in 2019 that the number of solar rooftop systems on homes, which typically rely on PV, will double worldwide to approximately 100 million by 2024. According to the IEA executive director, Fatih Birol, solar power is on its way to becoming the "new king" of the electricity sector.[10]

As the world goes solar, however, it has proven careless about considering the threats posed by climate impacts to solar generation. When solar power falls short, people and utilities are likely to turn to fossil fuels, be it coal- and gas-powered power plants or diesel-fueled generators, to fill the gaps. Professor Gary Rosengarten at

RMIT University in Australia termed this the "snowball effect," when climate-worsened events reduce the reliability of clean energy, thus increasing the reliance on more dependable fossil fuel–generated power. Another word for this is "malmitigation."[11] The snowball effect has already smacked California and Australia, two areas in the world deeply affected by climate-worsened fires.

Both regions have made major commitments to generating electricity from the sun. By 2018, solar power provided almost 20 percent of California's power (Figure 6.2). Some of the electricity came from huge solar arrays, like the Topaz Solar Farm with 8.4 million solar panels occupying over seven square miles (18.1 square kilometers) in the middle of the state. California also requires that all new homes include rooftop solar systems. Meanwhile,

FIGURE 6.2. Topaz Solar Farm, located in San Luis Obispo County, California, generates enough electricity to service more than 180,000 California homes at a cost far below coal and natural gas.

Source: Wikimedia, 'Solar Panels at Topaz Solar 7.'

Australia has the highest number of businesses and houses with rooftop solar panels anywhere in the world, soaring from 35,000 in 2010 to more than two million in 2019.

But as we saw in Chapter 5, drier, hotter conditions in California and Australia (and other parts of the world as well) have stoked bigger wildfires. The dirty smoke from those fires presents a challenge to solar power generation. It deposits ash and dust on the PV panels and prevents sunlight from reaching their cells, cutting electricity production. Indeed, in California, wildfires drove at least a 13 percent decline in solar generation on the grid during two weeks in September 2020, with some systems producing no solar power at all. In Australia, the bushfire-driven decline proved even more dramatic—as much as 45 percent in Sydney and Canberra on days with heavy smoke. In both locations, the electricity shortfalls forced utilities to ramp up or import electricity generated from coal or other fossil fuels, thus adding to carbon emissions.

And it's not just wildfires that will hamper solar power generation. Fluctuations in cloud cover can as well. A study published in 2020 by a team of researchers showed that higher temperatures increase the amount of particulates, aerosols, and moisture in the atmosphere, causing greater cloud formation and less solar radiation to reach the ground.[12] This phenomenon will be most pronounced in hot, arid regions like the American Southwest and the Middle East, the researchers determined—just where experts predict much greater expansion of solar power.

In the cases of wildfires or more clouds, energy system planners will need to consider whether the worsening impacts of climate change require new solar technologies that perform more effectively in challenging conditions. They might need to expand battery storage and build more transmissions lines, as well as seek other sources of energy to plan for the coming disruptions. They could additionally bridge the divide between adaptation and mitigation by reducing the threat of big wildfires through measures like climate-proofing electrical transmission, restricting development in fire-prone areas to reduce the risk of people igniting fires, and engaging in more controlled burns of forests and brush to reduce fuel loads that feed massive wildfires.

Planners need to account for climate impacts when it comes to wind power as well. In early 2021, an Arctic blast of freezing cold hit the state of Texas, costing people's lives and an estimated $90 billion in damages when the electric grid shut down. The power outages revealed that energy utility owners had failed to plan for cold extremes. As energy consultant Alison Silverstein described it, Texas "designed this entire grid for 'Ozzie and Harriet' weather; we are already facing 'Mad Max. . . . Everybody has always designed these systems looking in the rear-view mirror.'"[13] The cold revealed that this rear-view mirror approach was even used on newly-installed wind turbines when the turbines froze. Utility owners had not equipped the turbines with the tools that would allow them to operate in cold conditions.

As countries and communities seek to cut their emissions to slow global warming, they should also consider the ramifications of those choices in terms of the twin goals of reducing greenhouse gas emissions and building resilience to climate impacts. They should watch for predictable side effects such as the disruption of local ecosystems by the planting of massive amounts of trees to absorb carbon. As they opt for more wind or solar farms, they need to make sure that those facilities do not occupy the very land where the town might have to move in the future due to sea-level rise or increased riverine flooding, or that the process of land-clearing for construction does not lead to the loss of carbon-absorbing forests. Similarly, if nations build dams to produce hydropower, they should examine the downstream effects on water availability for communities dependent on the river's flow. And of course, all of these choices should account for the escalating impacts of climate.

In addition, communities and nations should search for creative win/win solutions like installing solar panels above crops to provide shade cover to plants and extra income to farmers or placing floating panels on reservoirs to reduce evaporation of fresh water in water-stressed regions. As former U.S. energy secretary Ernest Moniz remarked while speaking of possible shortages of electrical generating capacity going forward, "We need to do more in terms of looking at how the whole system fits together."[14] Nature-based solutions provide an early place to marry the goals of climate adaptation and mitigation.

Getting back to nature to mitigate and adapt

"Back to nature" is gaining traction as one of the most effective ways to converge the twin goals of adaptation and mitigation. Nature-based solutions offer the win/win of storing carbon and protecting against accelerating climate impacts. Nowhere is that dynamic more compelling than for coastal wetlands. Coastal wetlands—mangroves, seagrass meadows, and salt marshes—act as massive carbon sinks, earning them the moniker "blue carbon." In addition, they provide substantial protection against flooding, the most common natural hazard and one that rising temperatures worsen.

Mangrove forests, sturdy shrubs and trees with long, stilt-like roots, grow in the wet, salty, muddy soils of the tropics. Mangrove forests cover over 50,000 square miles (129,500 square kilometers) of the earth's coastlines, with the majority located in Southeast Asia. As waves wash through mangroves' tangled branches and roots, the waves lose energy. One hundred meters (328 feet) of mangroves reduces the potentially damaging force of waves by 13 to 66 percent. The natural buffer offered by mangroves prevented an estimated $1.5 billion in direct damages when Hurricane Irma pounded Florida in 2017. Wetlands saved Australians over $20 billion in storm damage in the five decades from 1967 to 2016. Scientists estimate that mangrove forests currently protect over five million people in the tropics from storms and prevent about $65 billion in damages worldwide annually.[15]

Mangroves have the added benefit of drawing carbon out of the atmosphere into the soil, offering a uniquely efficient storage capacity. They can sequester up to five times as much carbon as terrestrial forests. A 2018 study estimated that mangrove soil around the world held six billion tons of carbon in 2000, but between 2000 and 2015, the loss of mangrove forests due to, among other things, urban development, aquaculture, and mining caused the release of up to 122 million tons of carbon.[16] A 2020 study estimated that, in areas offering the most potential to buffer climate impacts, mangroves store about 10 percent of global fossil fuel emissions, even though they make up only 3 percent of the planet's forest cover.[17]

Vietnam has proven itself a leader in mangrove restoration. During what Vietnamese call the American War and what Americans call the Vietnam War from 1965 to 1969, the United States dropped bombs and sprayed chemical defoliants on mangrove forests in Ho Chi Minh City. As a result of the damage, riverbanks eroded and contaminants ruined the soil. An estimated 57 percent of the mangroves were destroyed. After the war, residents cut down even more mangroves to use in construction and as fuel.

Beginning in 1978, Vietnam began efforts to rehabilitate the forests, renaming the region Can Gio. In 1991, authorities established Can Gio as a coastal preserve under tight restrictions. Ho Chi Minh City has since emerged as the site of one of the largest restored mangrove forests in the world. The forests act as a carbon sink while providing a buffer against sea-level rise, which is occurring rapidly in the area. In the words of one official responsible for overseeing the mangrove reserve, "It serves as the city's 'lungs' by absorbing carbon dioxide and providing oxygen; it acts as the 'kidneys' by filtering wastewater pollution sent downstream from the city; and it is a 'green wall,' protecting the region from storms, typhoons, and tsunami coming from the East Sea."[18]

In the United Kingdom, the world's largest coastal habitat restoration project is offering protection against rising seas while serving as a carbon sink. Located on Wallasea Island in Essex, it involves the use of more than three million tons of clay excavated from a tunnel in London to build a 670-hectare (1,656-acre) stretch of mudflats, lagoons, and salt marshes.

Although one Dutch government scientist had floated the notion of building a 475-kilometer (295-mile) dam along the North Sea to protect this area and additional parts of Europe from flooding, the Wallasea project and others like it are a far less expensive approach. The Wallasea Island Wild Coast Project is expected to reduce wave height, capture soil that might otherwise erode away, serve as a natural buffer against the sea, and provide valuable habitat for rare and endangered wildlife, all while sequestering large amounts of carbon.

Wetlands restoration can also yield substantial economic gains from, for example, expanded fisheries and tourism. Research regarding one wetlands restoration project in the Philippines calculated that the direct benefits to the local community from restored mangroves greatly exceeded the restoration costs, even when the flood mitigation benefits were left out of the equation.[19] But to realize the promise of these natural solutions, the world must act quickly.

One important uncertainty about wetlands is whether they will manage to keep up with the rate of sea-level rise. Some research indicates that wetlands can gradually move inward along with the sea, but a 2020 study indicated that if seas rise faster than seven millimeters (0.27 inches) a year as opposed to approximately 3.4 millimeters (0.13 inches), wetlands simply can't keep up.[20] Since sea-level rise does not occur evenly like a basin filling with water, but rather unevenly around the globe, some areas are already seeing rates higher than this threshold. It takes time for the seagrasses and mangroves to take root and grow to provide the necessary buffer. If coastal communities delay their wetlands restoration efforts, they may discover that the sea-level rise has simply outpaced the ability of these natural systems to respond.

Paying greater policy attention to marrying the two goals of adaptation and mitigation produces bonuses in other areas as well. Take the agriculture and forestry sectors. In 2010, almost a quarter of all greenhouse gas emissions came from agriculture, forestry, and changes in land use and land cover through practices like deforestation. On the other hand, climate impacts such as longer droughts, heavier rainfall, and rising temperatures can slash agricultural production, threatening food security. "Climate-smart agriculture" has emerged as an effective approach for managing farmlands, fisheries, and forests. It promotes, for example, building resilience to climate-worsened threats like drought and pests while emphasizing use of less carbon-intensive farming methods to produce food. Climate-smart practices can avoid deforestation and thus increase carbon absorption.

Preserving and restoring healthy ecosystems that serve as carbon sinks and reduce climate impacts, like forests, grasslands, and marine areas, also promotes biodiversity, improves soil conservation, and decreases the vulnerability of surrounding communities to resource degradation. Yet policy alone will not guarantee success. It will take a substantial financial commitment as well. Financing the transitions to a carbon-neutral, resilient world will require financial innovation on an enormous scale. Fortunately, some countries are finding creative ways to raise funds that can satisfy both goals.

Financing both sides of the climate equation

To date, finance for adaptation has lagged behind that for mitigation in both developed and developing nations. This has led to a comparative dearth of expertise on the adaptation side of the ledger. The funding that has become available for adaptation has all too often gone into pilot or standalone programs, along with capacity building.[21] As the globe seeks to wed mitigation and adaptation strategies, it will need to find innovative solutions to leverage public and private finance to address both sides of the climate equation.

One potential way for nations to drive more investment in measures that jointly address adaptation and mitigation is through an environmental fiscal transfer. The transfer involves redistributing tax revenues among government entities to provide incentives for local community choices, for example, land-use choices. Fiscal transfers can broaden local acceptance of national policies, thus easing the implementation of large-scale projects that marry adaptation and mitigation.

Brazil pioneered the use of ecological fiscal transfer mechanisms to promote better land-use practices at the local level. Local Brazilian states had pleaded with the national government for reforms to address the distribution of value-added taxes (VATs), claiming that the fact that they had large swaths of protected national or state land left them fiscally disadvantaged since the VAT revenues only went to the states that generated the tax. Eventually the states took the matter to court and won. In 1991, the state of Paraná was the first to receive a share of tax revenues based on watershed and biodiversity protection measures. Since then, the ecological VAT program known as ICMS Ecológico has spread to most Brazilian states. Municipalities receive funds based on the amount of protected area they create as well as the quality of their public services. The better the environmental performance, the higher the rewards. Such a system could promote preservation of mangrove forests and other wetlands to double as buffers to storms and act as carbon sinks.

Mexico City likewise chose tax policy to achieve climate objectives. The city offers a 10 percent property tax reduction for installing green roofs in new and existing buildings. In the process, the city has become home to the largest single green roof in Latin America, covering 2,500 square meters (26,900 square feet). That roof and many others moderate temperatures year-round, making buildings warmer in the winter and cooler in the summer. By reducing the need for heating and cooling, the roofs cut energy consumption and greenhouse gas emissions. As added benefits, the plants in green roofs act as a carbon sink, reduce the heat island effect that occurs in cities, and help manage stormwater runoff (Figure 6.3).

Washington, DC, turned to debt financing to fund its climate solutions in 2016, offering the first-ever "environmental impact

FIGURE 6.3. The 2,500-square-meter green roof located in Mexico City atop INFONAVIT National Workers' Housing Fund Institute, the "lungs" of its community and the largest of its kind in Central America.
Source: Courtesy of CONViVe.

bond" in the United States. The city wanted to use green infra-structure measures like permeable pavement, landscaping, and green roofs to manage stormwater runoff, but needed capital to make it happen. The bonds take a "pay for success approach" to provide up-front funding to finance nature-based solutions. If the project performs well, the bonds pay investors a premium. On the other hand, if the project underperforms, the investors pay more. Designed to be scalable and replicable, environmental impact bonds quickly gained traction. In short order, the American cities of Atlanta, Baltimore, and Buffalo embraced the idea to fund their own green infrastructure projects.

The point is that as countries and communities work to solve their climate challenges, financing options need to exist to help achieve win/win solutions that fund both mitigation and adapta-tion activities.

Taking climate vows seriously

Nations can ill afford to ignore opportunities to marry efforts to cut emissions and build resilience going forward. One area that cries out for a conscious marriage of adaptation and mitigation princi-ples is future urbanization. Before the pandemic struck, countries around the planet had entered the biggest urban building boom in human history. More than half of the world's population now lives in cities. By 2060, two-thirds of the anticipated population of ten billion will live in urban areas. That level of expansion means that the world will double its existing building stock, adding almost 2.5 trillion square feet (230 billion square meters) of new floor area. That equates to adding a new Paris every week.

Designers and planners should ensure that new construction incorporates climate mitigation and adaptation considerations. Nowhere is this more important than when building an entirely new city, as Indonesia plans to. In 2019, Indonesian President Joko Widodo announced that the government would move the nation's populous capital city, Jakarta, on the island of Java, about 1,400 kilometers (870 miles) away to the island of Borneo, known as Kalimantan in Indonesia. Overcrowded and plagued with traffic congestion, Jakarta has become increasingly vulnerable to flooding. Subsidence due to poor urban planning has meant that 40 percent of the city now sits below sea level. The new city in Kalimantan would become home to an estimated 1.5 million public sector workers. Calling the new capital a "smart city in the forest," the government claims it will run on renewable energy. The city will additionally, according to the government, face less risk from climate-worsened flooding and other natural disasters like earthquakes and tsunamis.

Despite the benefits of the new location from an adaptation perspective, environmentalists fear that the move could release enormous amounts of carbon into the atmosphere. The area selected for the new capital includes both forestland and peatlands, a type of wetlands that sequesters large amounts of carbon. Clearing forests to build the new city would release tons of carbon and destroy a significant carbon sink. Similarly, draining peatlands to allow building to take place could release large amounts of greenhouse gas emissions. When peatlands become dry, they are highly susceptible to fires. Once ignited, peat fires can burn for months. Because peatlands are so dense with carbon, peat fires cause especially dirty smoke. Borneo already suffers from the acrid haze caused by such fires, which are sometimes set illegally to clear agricultural land. If Indonesia moves forward with this project, it needs to closely consider both sides of the climate equation. Poor land-use choices run the risk of exacerbating the climate challenges the nation already faces.

Focusing on the two sides of the climate equation is crucial not just in building new cities, but also in renovating existing ones. Existing cities and infrastructure systems, built at a time of abundant and cheap fossil fuels, continue to contribute to carbon emissions. It is estimated that about 60 percent of greenhouse gas emissions are hardwired into current infrastructure. Making sure that the 0.5 to 1 percent of all building stock that undergoes renovation annually incorporates adaptation and mitigation measures is essential to making progress on reducing climate risks.

But without policies or incentives in place to drive better decision-making, both renovations and new construction may fall woefully short in joining resilience and mitigation—to the detriment of the long-term safety of all. According to the fifth IPCC report, fewer than half of the G20 members integrated mitigation and adaptation considerations into infrastructure planning. Five failed to even mention either issue. This apparent omission leaves "considerable scope for G20 countries to heighten their efforts."[22] Governments, nongovernmental organizations (NGOs), academic institutions, and the private sector should find ways to improve building codes, building materials, construction techniques, and land-use practices, among other things, to give both adaptation and mitigation their due.

Marrying these two goals of climate adaptation and mitigation may not be appropriate or needed in every case, however. Even where linkage exists, the costs of pursuing both adaptation and mitigation could prove prohibitively expensive. Nor will it be easy to discern the second-, third-, and fourth-order consequences. Using scenarios in a systematic way could help weigh the risks and rewards of alternative approaches. As Cleo Paskal from the think tank Chatham House has urged, "We need to blow open the box on how complicated these problems are. We need as many different people involved and as many different sorts of solutions as possible."[23] Exploring the linkages will paint a more holistic picture of what's at stake, and in many cases, decision makers may discover the fortuitous surprise of a win/win solution.

CONCLUSION

Adaptation can no longer wait

We are bound together in a single garment of destiny.

—*Martin Luther King Jr.*

Over a century before the novel coronavirus began its spread in Wuhan, China, another type of contagion washed over that city—the contagion of revolution. On October 11, 1911, armed revolutionaries stormed a military garrison in Wuhan and immediately established a new Chinese government. Triggered in part by anger over corruption in the imperial court and foreign meddling in Chinese internal affairs, the Xinhai Revolution soon spread like a virus. In February 1912, the empress dowager abdicated her throne, marking the end of the Qing Dynasty and ending more than 2,000 years of feudal monarchy in China.

That same year, a family friend and the woman for whom I am named, Alice Chamberlayne, traveled to China at the age of 25. My family does not know her purpose or even her itinerary, but we do know that the voyage profoundly affected her worldview. A tall, imposing woman, she spent most of her life in Richmond, Virginia. When she died in 1968, she left a single, undated note referring to the journey she had made decades earlier to China.

Written in her large, even script, the paper carries the title, "World Affairs from the Kitchen Window." Shortly after her return from China, she wrote, she saw an old workman seated on her neighbor's kitchen stoop with his head in his hands. When she asked what was wrong, she learned that "rioting and trouble in China" had caused a tobacco factory in Richmond to close, costing the man his job. "Having just measured the distance between Virginia and China," Chamberlayne wrote, "I knew, from months of travel, the two places were pretty far apart." It "dawned" on her then that if what went on in China could affect her "own home and people, there wasn't any such thing as isolationism in the world." She had realized "in human & homely terms [that] the people of one country are bound up with the people of other countries." For this reason, she denounced the U.S. Congress's isolationism in opposing the League of Nations, the international peacekeeping organization that proved a failure in the run-up to World War II. Chamberlayne's reflection identified the dominant global change in her lifetime and the years after, the thickening sinews that knot the world into an ever-tighter ball. The image of the man with his head in his hands, she marveled, never left her.

***

Scientists mark 1880 as the start of modern global record-keeping for tracking climate data. Chamberlayne was born six years later in 1886. At that time, the accumulation of carbon emissions in the atmosphere stood at about 280 parts per million (ppm). Her birth came just a few decades after a railroad engineer in Titusville, Pennsylvania, discovered liquid petroleum, and the same year a German named Carl Benz applied for the first automobile patent.

Chamberlayne survived the 1918 flu pandemic and lived long enough to witness the post–World War II founding of the United Nations. But at the time of her death in 1968, scientific opinion had just begun to coalesce around the theory that carbon emissions from the burning of fossil fuels could dangerously accumulate in the atmosphere. Only a year before, two scientists, Syukuro Manabe and Richard Wetherald, had developed the first model capable of simulating the entire planet's climate. By then, the Mauna Loa Observatory in Hawaii had recorded a new high for carbon emissions in the atmosphere, 322 ppm.

A decade after Chamberlayne died, Robert M. White, a meteorologist by training who earned the moniker of the United States' "top weatherman" after serving as the advisor to five American presidents, predicted that climate change, in contrast to other environmental harms, could "utterly change our society and our civilization."[1] A decade after White's prediction, in 1988, scientists within Shell Oil Co. assessed that climate change could carry social, economic, and political consequences leading to upheaval powerful enough to rank as "the greatest in human history."[2] The scientists prophesied that it might take until the 2000s for sufficient evidence to exist to prove that the accumulation of carbon emissions forced unnatural changes to the climate. They urged their bosses, however, to consider Shell's role in causing climate change. "With the very long time scales involved," they apprised, "it would be tempting for society to wait until then before doing anything."[3]

On a sweltering day in June 1988, the warnings of a climate scientist finally landed center stage when NASA scientist James Hansen told the U.S. Congress that he was 99 percent certain that record temperatures resulted from the growing concentration of greenhouse gases and were not the result of natural variation. That same month, Mauna Loa Observatory measured a new record for carbon emissions, 353 ppm. Two years later, the UN Intergovernmental Panel on Climate Change issued its first statement that the accumulation of greenhouse gases would result in "a rise of global mean temperature . . . greater than any in man's history." True to White's prediction and Shell's warning, and probably Chamberlayne's disappointment, efforts by the world's nations to arrest the growth of harmful emissions have failed to halt the climb.

Thousands of summits, conferences, meetings, books, reports, white papers, editorials, and other exhortations warning of climate change calamity have not yet resulted in sufficient political will to act. When international representatives gathered in an airport hangar outside of Paris at the end of 2015, they had largely given up on the idea of setting binding global targets and timetables for emissions reduction. Instead, the 192 signatories to the Paris Agreement settled on the concept of voluntary reductions. The historic agreement recognized that to preserve any semblance of the climate that the planet has experienced since the dawn of human civilization, greenhouse gas emissions must plunge quickly. Carbon emissions by then hovered at 400 ppm. Five years later, during the pandemic, emissions topped 418 ppm.

According to 2020 estimates from the United Nations, the world is not on track to meet the Paris goal. Admittedly, as the pandemic wore on, a number of nations made ambitious pledges for becoming carbon neutral or net zero by 2050 to 2060, including, importantly, China and the United States, the two largest emitters. But even if the world's nations succeed in pulling the hand brake on emissions, which I fervently hope they do, we can no longer afford to ignore the climate change impacts already drubbing the planet.

There is no escaping the need to adapt.

Increasingly, the academic literature has called for "transformational change" to address the climate crisis. Given the systemic nature of climate risk, with the potential to cascade through ecological, social, and financial systems, the appeal of transformational change is easy to discern. Transformational change can overthrow entrenched social systems, as the abolitionist movement to end slavery did in the 19th century, or spawn new economic and cultural configurations, as the Industrial Revolution did. Transformational change may well hang ripening on a tree like a piece of fruit. But to date the world has seemingly failed at building the necessary ladder to reach it. Predicting when a transformational moment might arrive for climate change is about as easy as predicting a hurricane in Florida based on the flap of a butterfly's wings in Indonesia.

And with climate change, there is never going to be a vaccine.

So how do we deal with a future that promises more climate-worsened catastrophes?

Do what humans do well: solve the immediate problem before us. When I supervised the white-collar crime unit at the U.S. Attorney's Office in Los Angeles, a young, eager prosecutor came to me for advice. Assigned to a sprawling fraud investigation, he asked, "How can I possibly know where to begin?" He lamented, "There's just too much material to cut through to make any sense of how to move forward." The important thing, I told him, was to get started. Pick a place and begin. I assured him that he could keep looking ahead, but when our headlights only reach so far, we must make progress where we can. That advice, I believe, holds true for policymakers and decision makers seeking to move forward on climate change. No one person or business or government can see beyond our collective headlights to predict with confidence what humans will choose to do. But each of us can get started, finding the path that helps solve the problem before us, while remaining flexible and ready should the moment of transformational change arrive.

Solutions exist for solving the problems at our feet, as do ideas for how to scan the horizon for what's ahead. Since climate change affects everything, there is no shortage of actions that governments, communities, nongovernmental organizations (NGOs), philanthropies, businesses, and individuals can take now to reduce climate risk. Creating a national plan that identifies roles for all these stakeholders will promote a coordinated approach across all sectors and divisions of government and society. A national plan, among other things, can improve disaster response, help prioritize public investments in adaptation, pinpoint critical research needs, and, importantly, identify markers for measuring progress.

To address the fundamental challenge of preparing for climate change, namely that the past is no longer a safe guide for the future, governments, in collaboration with affected stakeholders, should assist communities, businesses, and households in planning for a world where physical impacts will rapidly intensify. This requires developing climate-resilient building codes and standards, crafting land-use policies that remove people from harm's way, and prioritizing nature-based solutions, such as mangrove forests and salt marshes to buffer sea-level rise. These decisions lie at the heart of adaptation.

To repair the cracks in emergency response efforts exposed by the pandemic, governments, communities, businesses, NGOs, and individuals should beef up emergency preparedness by building stockpiles, fattening supply chains, creating surge workforces, and identifying ways for people to obtain aid immediately before and after disaster strikes. The pandemic also made clear that disasters don't stop at borders.

Countries, communities, businesses, philanthropies, NGOs, and households can work across boundaries to reduce climate risk. The borders erected throughout history often impede adaptation since climate change brings borderless disasters. Greater cooperation at all levels is required to reduce cross-border threats. This means, for example, that communities upstream should consider the water needs of their downstream neighbors as they work to address climate-fueled water scarcity. Reinvigorating multilateralism could markedly boost these efforts, particularly when it comes to internationally shared river basins as well as changing fisheries. Stronger multilateral governance could further provide much-needed oversight to human attempts to intervene in the climatic system, imposing order before geoengineering experiments inadvertently cause grave harm.

The pandemic also unveiled the crushing burden that catastrophe imposes on the most vulnerable. It may well have placed the UN's Sustainable Development Goals out of reach. COVID-19's toll on women, children, the poor, older people, people with disabilities, and the marginalized broadcast an urgent signal to build programs attentive to the needs of those most at risk. Countries, NGOs, philanthropies, communities, and businesses should start that work now. Good places to start are to make sure that the most vulnerable people have a seat at the table for adaptation planning and that any adaptation efforts address the inequality suffered by women and girls.

To jumpstart understanding of future climate disruptions, nations can require companies to scour their operations and assets for possible climate risk. Requiring disclosure by businesses will help the businesses themselves, as well as communities and investors, to better understand the ballooning threats. By making sure that past disasters do not slip from memory, governments, communities, and businesses can help encourage people to prepare, reminding them that it's only a matter of time before calamity strikes again.

Going forward, decision makers can marry mitigation and adaptation strategies so that they work in concert, not at odds, with one other. For example, governments, communities, and households should make sure that solar energy provides the necessary power when more soot from wildfires and changes in cloud cover block the sun from hitting the photovoltaic panels. And before they turn to desalination, they should insist on greater conservation of water. Meeting the climate challenge requires all of us to start rowing in the same direction in ways we haven't before.

The explorer Ernest Shackleton once observed, "Optimism is true moral courage." If anyone knew about courage, he did. Shackleton set out on his ship *Endurance* to cross Antarctica from sea to sea in August 1914, but disaster struck when ice in the Weddell Sea surrounded the vessel in January 1915. Eventually, the ice closed around the *Endurance*, crushing it from stem to stern 11 months later. Shackleton and his crew abandoned the vessel in search of help. Their journey to safety, with no loss of life, is one of the most remarkable adventure tales of all time. It's also a story of leadership in crisis, taught in executive training and business schools today. Fighting climate change may very well require the moral courage of optimism as we pursue a better future. My aim is to supply the insights needed to fortify that courage.

NOTES

Introduction

1. David Wallace-Wells, "After Alarmism: The War on Climate Denial Has Been Won. And That's Not the Only Good News," *New York Magazine*, January 19, 2021, https://nymag.com/intelligencer/article/climate-change-after-pandemic.html.
2. I sit on the board of Munich Re America and related U.S. subsidiaries of the Munich Re Group.
3. Luke Gallin, "Global Insured Nat Cat Losses up 44% in 2020 to $82bn: Munich Re," Reinsurance News, January 7, 2021, https://www.reinsurancene.ws/global-insured-nat-cat-losses-up-44-in-2020-to-82bn-munich-re/. I sit on the boards of the domestic subsidiaries of the Munich Re Group.
4. Robert McSweeney, "Climate Change Set to Increase Extreme Weather Risk to UK Population," Carbon Brief, last modified November 27, 2014, https://www.carbonbrief.org/climate-change-set-to-increase-extreme-weather-risk-to-uk-population.
5. Rico Kongsager, "Linking Climate Change Adaptation and Mitigation: A Review with Evidence from the Land-Use Sectors," *Land* 7, no. 158 (2018), https://doi.org/10.3390/land7040158.
6. Yoly Gutierrez, "Climate Change: Time to Combine Mitigation and Adaptation," CGIAR Research Programme on Forests, Trees, and Agroforestry, last modified January 15, 2017, https://www.foreststreesagroforestry.org/news-article/climate-change-time-to-combine-mitigation-and-adaptation/.
7. António Guterres, "Remarks to the Climate Adaptation Summit," January 25, 2021, UN Headquarters, New York City, United States, un.org/sg/en/content/sg/speeches/2021-01-25/remarks-climate-adaptation-summit.
8. Barbara Buchner et al., *Global Landscape of Climate Finance 2019* (London: Climate Policy Initiative, 2019), https://www.climatepolicyinitiative.org/wp-content/uploads/2019/11/2019-Global-Landscape-of-Climate-Finance.pdf.

9. Dellmuth Lisa Maria, Gustafsson Maria-Therese, and Kural Ece, "Global Adaptation Governance: Explaining the Governance Responses of International Organizations to New Issue Linkages," *Environmental Science and Policy* 114 (2020): 204–215, https://www.sciencedirect.com/science/article/pii/S1462901120303907?dgcid=rss_sd_all.

Chapter 1

1. Susan E. Rice, "The Government Has Failed on Coronavirus, But There Is Still Time," *New York Times*, March 13, 2020, https://www.nytimes.com/2020/03/13/opinion/corona-virus-trump-susan-rice.html?referringSource=articleShare.

2. David Quammen, "Why Weren't We Ready for the Coronavirus?," *New Yorker*, May 4, 2020, https://www.newyorker.com/magazine/2020/05/11/why-werent-we-ready-for-the-coronavirus.

3. Chi Xu et al. "Future of the Human Climate Niche," *Proceedings of the National Academy of Sciences of the United States* 117, no. 21 (2020): 11350–11355, https://doi.org/10.1073/pnas.1910114117.

4. Allie Goldstein et al. "The Private Sector's Climate Change Risk and Adaptation Blind Spots," *Nature Climate Change* 9 (2019), 18–25, https://doi.org/10.1038/s41558-018-0340-5.

5. CNA Military Advisory Board, *National Security and the Accelerating Risks of Climate Change* (Alexandria, VA: CNA Corporation, 2014), https://www.cna.org/CNA_files/pdf/MAB_5-8-14.pdf.

6. Peter Matanle, Joel Littler, and Oliver Slay, "Imagining Disasters in the Era of Climate Change: Is Japan's Seawall a New Maginot Line?," *Asia-Pacific Journal* 17, no. 13 (2019), https://apjjf.org/2019/13/Matanle.html.

7. *Fourth National Climate Assessment, Volume II: Impacts, Risks, and Adaptation in the United States* (Washington, DC: U.S. Global Change Research Program, 2018), https://nca2018.globalchange.gov/.

8. *Policy Guide on Planning and Climate Change* (Chicago: American Planning Association, 20011), https://planning-org-uploaded-media.s3.amazonaws.com/legacy_resources/policy/guides/pdf/climatechange.pdf.

9. Sara Mehryar and Swenja Surminski, "National Laws for Enhancing Flood Resilience in the Context of Climate Change: Potential and Shortcomings," *Climate Policy* 13, no. 4 (2020), https://www.tandfonline.com/doi/full/10.1080/14693062.2020.1808439.

10. Marta Olazabal, "A Cross-Scale Worldwide Analysis of Coastal Adaptation Planning," *Environmental Research Letters* 14, no. 12 (2019), https://iopscience.iop.org/article/10.1088/1748-9326/ab5532.

11. Global Commission on Adaptation, *Adapt Now: A Global Call for Leadership on Climate Resilience* (Washington, DC: World Resources Institute, 2019), https://cdn.gca.org/assets/2019-09/GlobalCommission_Report_FINAL.pdf.

12. Global Commission on Adaptation, *Adapt Now: A Global Call for Leadership on Climate Resilience* (Washington, DC: World Resources Institute, 2019), https://cdn.gca.org/assets/2019-09/GlobalCommission_Report_FINAL.pdf.

13. "U.S. Climate Resilience Toolkit," U.S. Global Change Research Program, last modified March 25, 2020, https://toolkit.climate.gov/topics/coastal/storm-surge.

14. *Flooded Future: Global Vulnerability to Sea Level Rise Worse Than Previously Understood* (Princeton, NJ: Climate Central, 2019), https://www.climatecentral.org/news/report-flooded-future-global-vulnerability-to-sea-level-rise-worse-than-previously-understood.

15. *Flooded Future: Global Vulnerability to Sea Level Rise Worse Than Previously Understood* (Princeton, NJ: Climate Central, 2019), https://www.climatecentral.org/news/report-flooded-future-global-vulnerability-to-sea-level-rise-worse-than-previously-understood.

16. E&E News, "Technology: Investors Gauge Future Climate Risks with Satellite Imaging," *ClimateWire* 10, no. 9 (2020), https://www.eenews.net/climatewire/2020/10/23/stories/1063716893?utm_campaign=edition&utm_medium=email&utm_source=eenews%3Aclimatewire.

17. J. Egeland and J. Chissano, *Climate Knowledge for Action: A Global Framework for Climate Services* (Geneva: World Meteorological Organization, 2011), https://gfcs.wmo.int/sites/default/files/FAQ/HLT/HLT_FAQ_en.pdf.

18. C. D. Hewitt et al. "Making Society Climate Resilient: International Progress under the Global Framework for Climate Services," *Bulletin of the American Meteorological Society* 101, no. 2 (2020): E237–E252, https://doi.org/10.1175/BAMS-D-18-0211.1.

19. Jesse Keenan, "A Climate Intelligence Arms Race in the Financial Markets," *Science* 365, no. 6459 (2019), https://science.sciencemag.org/content/365/6459/1240.abstract.

20. Paula Jarzabkowski et al. "Insurance for Climate Adaptation: Opportunities and Limitations," Global Commission on Adaptation,

July 2019, https://cdn.gca.org/assets/2019-07/Insurance%20for%20cli-mate%20adaptation_Opportunities%20and%20Limitations.pdf.

21. David D. Kirkpatrick, Matt Apuzzo, and Selam Gebrekidan, "Europe Said It Was Pandemic-Ready. Pride Was Its Downfall," *New York Times*, July 20, 2020, https://www.nytimes.com/2020/07/20/world/europe/coronavirus-mistakes-france-uk-italy.html?referringSource=articleShare.

22. David D. Evans, Cody Webb, and Eric J. Xu, "Wildfire Catastrophe Models Could Spark the Changes California Needs," Milliman, October 28, 2019, https://www.milliman.com/en/insight/wildfire-catastrophe-models-could-spark-the-changes-california-needs.

23. David D. Evans, Cody Webb, and Eric J. Xu, "Wildfire Catastrophe Models Could Spark the Changes California Needs," Milliman, October 28, 2019, https://www.milliman.com/en/insight/wildfire-catastrophe-models-could-spark-the-changes-california-needs.

24. Harvey V. Fineberg, "Preparing for Avian Influenza: Lessons from the 'Swine Flu Affair,'" *Journal of Infectious Diseases* 197, no. 1 (2008): 14–18, https://doi.org/10.1086/524989.

25. Dan Diamond and Nahal Toosi, "Trump Team Failed to Follow NSC's Pandemic Playbook," *Politico*, March 25, 2020, https://www.politico.com/news/2020/03/25/trump-coronavirus-national-security-council-149285.

26. Tom Inglesby and Eric Toner, "Our Lack of Pandemic Preparedness Could Prove Deadly," *Washington Post*, September 19, 2018, https://www.washingtonpost.com/opinions/our-lack-of-pandemic-preparedness-could-prove-deadly/2018/09/19/0d7b235c-b13e-11e8-a20b-5f4f84429666_story.html.

27. Matthew Belvedere, "Trump Says He Trusts China's Xi on Coronavirus and the US Has It 'Totally under Control,'" *CNBC*, January 22, 2020, https://www.cnbc.com/2020/01/22/trump-on-coronavirus-from-china-we-have-it-totally-under-control.html.

28. Katherine Eban, "How Jared Kushner's Secret Testing Plan 'Went Poof into Thin Air,'" *Vanity Fair*, July 30, 2020, https://www.vanityfair.com/news/2020/07/how-jared-kushners-secret-testing-plan-went-poof-into-thin-air.

29. Yasmeen Abutaleb et al. "70 Days of Denial, Delays and Dysfunction," *Washington Post*, April 5, 2020, https://www.washingtonpost.com/national-security/2020/04/04/coronavirus-government-dysfunction/?arc404=true&itid=lk_inline_manual_7.

30. Chad Terhune et al. "Special Report: How Korea Trounced U.S. in Race to Test People for Coronavirus," *Reuters*, March 18, 2020, https://www.reuters.com/article/us-health-coronavirus-testing-specialrep/special-report-how-korea-trounced-u-s-in-race-to-test-people-for-coronavirus-idUSKBN2153BW.

31. "How the World's Premier Public-Health Agency Was Handcuffed," *The Economist*, May 23, 2020, https://www.economist.com/node/21786465?frsc=dg%7Ce.

32. John Jullens, "How CEOs Can Respond to COVID-19 and Build a Resilient Future," *KPMG*, 2020, https://advisory.kpmg.us/articles/2020/covid-19-ceos-future.html.

33. Ryan Smith, "CEOs Admit They Aren't Prepared for Climate Change – FM Global Report," Insurance Business America, June 18, 2020, https://www.insurancebusinessmag.com/us/news/catastrophe/ceos-admit-they-arent-prepared-for-climate-change--fm-global-report-225583.aspx.

34. Dan Diamond, "Inside America's 2-Decade Failure to Prepare for Coronavirus," *Politico*, March 11, 2020, https://www.politico.com/news/magazine/2020/04/11/america-two-decade-failure-prepare-coronavirus-179574.

Chapter 2

1. "Somalia Struggles after Worst Flooding in Recent History," E&E News, November 15, 2019, https://www.eenews.net/climatewire/2019/11/15/stories/1061551557.

2. Rebecca Heisler, "Meteorologists Can't Keep Up with Climate Change in Mozambique," NPR, December 11, 2019, https://www.npr.org/sections/goatsandsoda/2019/12/11/782918005/meteorologists-cant-keep-up-with-climate-change-in-mozambique.

3. Stockholm University, "Climate Changes Make Some Aspects of Weather Forecasting Increasingly Difficult," Science X, March 25, 2019, https://phys.org/news/2019-03-climate-aspects-weather-increasingly-difficult.html.

4. Global Commission on Adaptation, *Adapt Now: A Global Call for Leadership on Climate Resilience* (Washington, DC: World Resources Institute, 2019), https://gca.org/global-commission-on-adaptation/report.

5. "'Humanitarian History' Made as Uganda Red Cross Launches Forecast-Based Financing For Real," *Climate Centre*, November 15, 2015,

https://www.climatecentre.org/news/657/a-humanitarian-historya-made-as-uganda-red-cross-launches-forecast-based-financing-for-real.

6. Emily Wilkinson et al., *Forecasting Hazards, Averting Disasters: Implementing Forecast-Based Early Action at Scale* (London: Overseas Development Institute, 2018), https://www.odi.org/sites/odi.org.uk/files/resource-documents/12104.pdf.

7. *Forecast-Based Financing in Nepal: A Return on Investment Study* (Rome: World Food Programme, 2019), https://docs.wfp.org/api/documents/WFP-0000108408/download/.

8. U.S. Congress, Senate, Committee on Homeland Security and Governmental Affairs, *Evaluating the Federal Government's Procurement and Distribution Strategies in Response to the COVID-19 Pandemic*, 116th Cong., 2nd sess., 2020, https://www.hsgac.senate.gov/evaluating-the-federal-governments-procurement-and-distribution-strategies-in-response-to-the-covid-19-pandemic.

9. Emmanuel Macron, interview by Roula Khalaf and Victor Mallet, April 17, 2020, transcript, *Financial Times*, https://www.ft.com/content/317b4f61-672e-4c4b-b816-71e0ff63cab2.

10. Vijay Vaitheeswaran, "Supply Chains Are Undergoing a Dramatic Transformation," *The Economist*, July 11, 2019, https://www.economist.com/special-report/2019/07/11/supply-chains-are-undergoing-a-dramatic-transformation.

11. UNDRR, *Global Assessment Report on Disaster Risk Reduction* (Geneva: United Nations Office for Disaster Risk Reduction, 2019), https://www.undrr.org/publication/global-assessment-report-disaster-risk-reduction-2019.

12. Nicoletta Batini, James Lornax, and Divya Mehra, "Why Sustainable Food Systems Are Needed in a Post-COVID World," International Monetary Fund, July 14, 2020, https://blogs.imf.org/2020/07/14/why-sustainable-food-systems-are-needed-in-a-post-covid-world/?utm_medium=email&utm_source=govdelivery.

13. Kaamil Ahmed, "Hunger Could Kill Millions More Than Covid-19, Warns Oxfam," *The Guardian*, July 9, 2020, https://www.theguardian.com/global-development/2020/jul/09/hunger-could-kill-millions-more-than-covid-19-warns-oxfam?CMP=oth_b-aplnews_d-1.

14. Kaamil Ahmed, "Hunger Could Kill Millions More Than Covid-19, Warns Oxfam," *The Guardian*, July 9, 2020, https://www.theguardian.com/global-development/2020/jul/09/

hunger-could-kill-millions-more-than-covid-19-warns-oxfam?CMP=
oth_b-aplnews_d-1.

15. Miroslav Trnka et al., "Mitigation Efforts Will Not Fully Alleviate
the Increase in Water Scarcity Occurrence Probability in Wheat-
Producing Areas," *Science Advances* 5, no. 9 (2019), https://advances.
sciencemag.org/content/5/9/eaau2406?rss=1.

16. Global Commission on Adaptation, *Adapt Now: A Global Call for Leadership on Climate Resilience* (Washington, DC: World Resources Institute,
2019), https://cdn.gca.org/assets/2019-09/GlobalCommission_
Report_FINAL.pdf.

17. Christina Anderson and Henrik Pryser Libell, "Finland, 'Prepper
Nation of the Nordics,' Isn't Worried about Masks," *New York Times*,
April 5, 2020, https://www.nytimes.com/2020/04/05/world/europe/
coronavirus-finland-masks.html.

18. Franck Galtier, *Can ECOWAS Regional Reserve Project Improve the
Management of Food Crises in West Africa? Case Study Report for the ASiST
Study "Which Role for Food Reserve in Improving Food Security in Developing
Countries?"* (Apsley: DAI, 2016), https://agritrop.cirad.fr/587252/1/
ECOWAS%20RR%20Report.pdf.

19. Dr. Michael Taylor, "Climate Change in the Caribbean – Learning
Lessons from Irma and Maria," *The Guardian*, October 6, 2017,
https://www.theguardian.com/environment/2017/oct/06/
climate-change-in-the-caribbean-learning-lessons-from-irma-
and-maria.

20. Gregory Korte, "White House Puts Fire Chiefs on Front Lines of Climate
Change," USA Today, November 9, 2015, https://www.usatoday.com/
story/news/politics/2015/11/09/white-house-puts-fire-chiefs-front-
lines-climate-change/75476480/.

21. Joseph Serna, "Growing Fire Threat Makes California Departments
Reluctant to Help Each Other," *Los Angeles Times*, November 9, 2019,
https://www.latimes.com/california/story/2019-11-09/california-
wildfire-mutual-aid-fire-departments.

22. Thomas Franck, "Firefighters Will Attack Blazes Quickly to Avoid
Coronavirus," E&E News, June 11, 2020, https://www.eenews.net/
climatewire/2020/06/11/stories/1063366291?utm_medium=email&utm_
source=eenews%3Aclimatewire&utm_campaign=edition%2BiZ%2B%2
FftFV%2B2LxUfHtN5bxJQ%3D%3D.

23. "New NSTB Publication Examines 30-Year History of Transportation
Safety Improvements," *Brotherhood of Locomotive Engineers and*

Trainmen, March 16, 2005, https://www.ble-t.org/pr/news/headline. asp?id=13055.

24. Ahmed Nagi, "Yemen and Coronavirus," in *Coronavirus in Conflict Zones: A Sobering Landscape*, ed. Jarrett Blanc and Frances Brown (Washington, DC: Carnegie Endowment for International Peace, 2020), https:// carnegie-mec.org/2020/04/14/yemen-and-coronavirus-pub-81534.

25. Kevin Sieff, Susannah George, and Kareem Fahim, "Now Joining the Fight against Coronavirus: The World's Armed Rebels, Drug Cartels and Gangs," *Washington Post*, April 14, 2020, https://www.washingtonpost. com/world/the_americas/coronavirus-taliban-ms-13-drug-cartels-gangs/2020/04/13/83aa07ac-79c2-11ea-a311-adb134471a9_story.html.

26. Elizabeth Ferris, "Natural Disasters, Conflict, and Human Rights: Tracing the Connections," *Brookings Institution*, March 3, 2010, https://www.brookings.edu/on-the-record/natural-disasters-conflict-and-human-rights-tracing-the-connections/.

27. "East Pakistani Leaders Assail Yahya on Cyclone Relief," *New York Times*, November 24, 1970, https://www.nytimes.com/1970/11/24/archives/east-pakistani-leaders-assail-yahya-on-cyclone-relief.html.

28. Jamie Dettmer, "Mafia Gangs in Italy Poised to Profit from Coronavirus," *VOA News*, April 3, 2020, https://www.voanews.com/science-health/coronavirus-outbreak/mafia-gangs-italy-poised-profit-coronavirus.

29. Marco Tedesco, "The Virus and the Cyclone: The Tragedy in India and Bangladesh Is Double," State of the Planet, Columbia University Earth Institute, May 28, 2020, https://blogs.ei.columbia.edu/2020/05/28/coronavirus-cyclone-amphan/.

30. "Recovery Begins after Deadly Cyclone Amphan Ravages India, Bangladesh," France 24, May 22, 2020, https://www.france24.com/en/20200522-recovery-begins-after-deadly-cyclone-amphan-ravages-india-bangladesh.

31. AFP, "Cyclone Toll Hits 95 as Bangladesh and India Start Mopping Up," *Economic Times*, May 22, 2020, https://economictimes.indiatimes.com/news/politics-and-nation/cyclone-toll-hits-95-as-bangladesh-and-india-start-mopping-up/articleshow/75882197.cms.

Chapter 3

1. James Fallows, "The 3 Weeks That Changed Everything," *The Atlantic*, June 29, 2020, https://www.theatlantic.com/politics/archive/2020/06/how-white-house-coronavirus-response-went-wrong/613591/.

2. Kate Wheeling, "Modeling the Cascading Infrastructure Impacts of Climate Change," *Eos: Earth and Space Science News*, October 19, 2020, https://eos.org/research-spotlights/modeling-the-cascading-infrastructure-impacts-of-climate-change.

3. South Carolina Floodwater Commission, *South Carolina Floodwater Commission Report* (Columbia: South Carolina Floodwater Commission, 2019), https://governor.sc.gov/sites/default/files/Documents/Floodwater%20Commission/SCFWC%20Report.pdf.

4. Michelle Ann Miller et al., *Crossing Borders: Governing Environmental Disasters in a Global Urban Age in Asia and the Pacific*, ed. Michelle Ann Miller, Michael Douglass, and Matthias Garschagen (Singapore: Springer Singapore, 2018), https://books.apple.com/us/book/crossing-borders/id1322135060.

5. Aaron Wolf, "Comprehensive Report of World's Transboundary Water Basins Finds Hotspots of Risk," Oregon State University, May 2, 2016, https://today.oregonstate.edu/archives/2016/may/comprehensive-report-world's-transboundary-water-basins-finds-hotspots-risk.

6. Ajar Sharma, Keith Hipel, and Vanessa Schweizer, "Strategic Insights into the Cauvery River Dispute in India," *Sustainability* 12, no. 4 (2020), https://www.researchgate.net/publication/339184358_Strategic_Insights_into_the_Cauvery_River_Dispute_in_India.

7. Simon Denyer and Chris Mooney, "2°C Beyond the Limit: The Climate Chain Reaction That Threatens the Heart of the Pacific," *Washington Post*, November 12, 2019, https://www.washingtonpost.com/graphics/2019/world/climate-environment/climate-change-japan-pacific-sea-salmon-ice-loss/.

8. Jonathan Lenoir et al., "Species Better Track Climate Warming in the Oceans Than on Land," *Nature Ecology & Evolution* 4 (2020): 1044–1059, https://doi.org/10.1038/s41559-020-1198-2.

9. Kisei R. Tanaka et al., "North Pacific Warming Shifts the Juvenile Range of a Marine Apex Predator," *Scientific Reports* 11, no. 3373 (2021), https://doi.org/10.1038/s41598-021-82424-9.

10. Kimberly L. Oremus et al., "Governance Challenges for Tropical Nations Losing Fish Species Due to Climate Change," *Nature Sustainability* 3 (2020): 277–280, https://doi.org/10.1038/s41893-020-0476-y.

11. "Rules and Consequences: How to Improve International Fisheries," Pew Charitable Trusts, July 20, 2020, https://www.pewtrusts.org/en/research-and-analysis/issue-briefs/2020/07/rules-and-consequences-how-to-improve-international-fisheries.

12. Malin L. Pinsky et al., "Preparing Ocean Governance for Species on the Move," *Science* 360, no. 6394 (2018): 1189–1191, https://science.sciencemag.org/content/360/6394/1189?referringSource=articleShare.

13. I serve on the board of trustees for the Environmental Defense Fund.

14. Franck Galtier, *Can ECOWAS Regional Reserve Project Improve the Management of Food Crises in West Africa? Case Study Report for the ASiST Study "Which Role for Food Reserve in Improving Food Security in Developing Countries?"* (Apsley: DAI, 2016), https://agritrop.cirad.fr/587252/1/ECOWAS%20RR%20Report.pdf.

15. "Invest in Early Warning to Deliver Climate Adaptation," World Meteorological Organization, September 9, 2019, https://public.wmo.int/en/media/news/invest-early-warning-deliver-climate-adaptation.

16. "How the African Risk Capacity Works," African Risk Capacity, last accessed April 2020, https://www.africanriskcapacity.org/about/how-arc-works/.

17. Daniel J. Clarke and Ruth Vargas Hill, *Cost-Benefit Analysis of the African Risk Capacity Facility* (Washington, DC: International Food Policy Research Institute, 2013), https://papers.ssrn.com/sol3/papers.cfm?abstract_id=2343159.

18. "CCRIF Announces Final Payout Numbers of US$12.8 Million to the Bahamas Following Hurricane Dorian," Caribbean Catastrophe Risk Insurance Facility, September 26, 2019, https://reliefweb.int/sites/reliefweb.int/files/resources/Press_Release_CCRIF_Makes_Full_Payouts_to_Bahamas_September_26_2019.pdf.

19. Karen McVeigh, "Coronavirus: World Bank's Pandemic Bonds Scheme Accused of 'Waiting for People to Die,'" Bhekisisa Centre for Health Journalism, March 2, 2020, https://bhekisisa.org/health-news-south-africa/2020-03-02-coronavirus-world-banks-500m-pandemic-bonds-sche.

20. Michael Igoe, "World Bank Pandemic Facility 'an Embarrassing Mistake,' Says Former Chief Economist," Devex, April 12, 2019, https://www.devex.com/news/world-bank-pandemic-facility-an-embarrassing-mistake-says-former-chief-economist-94697.

21. William A. Kandel, *Unaccompanied Alien Children: An Overview*, CRS Report No. R43599 (Washington, DC: Congressional Research Service, 2019), https://www.everycrsreport.com/reports/R43599.html.

22. WFP Regional Bureau for Latin America and the Caribbean, *Food Security and Emigration* (City of Knowledge: World Food Programme,

2017), https://reliefweb.int/sites/reliefweb.int/files/resources/WFP-0000022124.pdf.

23. Abrahm Lustgarten, "The Great Climate Migration," *New York Times*, July 23, 2020, https://www.nytimes.com/interactive/2020/07/23/magazine/climate-migration.html.

24. More than 30 African nations have formally defined duties and responsibilities for protecting and assisting internally displaced persons through the African Union Convention for the Protection and Assistance of Internally Displaced Persons in Africa (the Kampala Convention), which came into effect in 2012.

25. Robert E. Rubin and David Miliband, "Borders Won't Protect Your Country from Coronavirus," *New York Times*, July 6, 2020, https://www.nytimes.com/2020/07/06/opinion/coronavirus-aid-miliband.html?referringSource=articleShare.

26. Natalie Kitroeff, "2 Hurricanes Devastated Central America. Will the Ruin Spur a Migration Wave?," *New York Times*, December 4, 2020, https://www.nytimes.com/2020/12/04/world/americas/guatemala-hurricanes-mudslide-migration.html?referringSource=articleShare.

27. Alison Rourke, "Global Report: Red Cross Warns of Big Post-Covid-19 Migration as WHO Hits Back at US," *The Guardian*, July 23, 2020, https://www.theguardian.com/world/2020/jul/24/global-report-red-cross-warns-of-big-post-covid-19-migration-as-who-hits-back-at-us?CMP=oth_b-aplnews_d-1.

28. Michael Soussan, "The Case for Passports for the World's Refugees," *Pacific Standard Magazine*, last modified May 3, 2017, https://psmag.com/news/the-case-for-passports-for-the-worlds-refugees.

29. John M. Barry, *The Great Influenza: The Story of the Deadliest Pandemic in History* (New York: Penguin Publishing Group, 2005), 456.

30. Geoffrey D. Dabelko et al., eds., *Backdraft: The Conflict Potential of Climate Change Adaptation and Mitigation* (Washington, DC: Woodrow Wilson International Center for Scholars, 2013), https://issuu.com/ecspwwc/docs/ecsp_report_14_2_backdraft.

Chapter 4

1. *The Sustainable Development Goals Report 2019* (New York: United Nations, 2019), https://unstats.un.org/sdgs/report/2019/The-Sustainable-Development-Goals-Report-2019.pdf.

2. Samuel Freije-Rodriguez et al., *Poverty and Shared Prosperity 2020: Reversals of Fortune* (Washington, DC: World Bank Group, 2020), https://openknowledge.worldbank.org/bitstream/handle/10986/34496/9781464816024.pdf.

3. Heather Randell and Clark Gray, "Climate Variability and Educational Attainment: Evidence from Rural Ethiopia," *Global Environmental Change* 41 (2016), https://www.sesync.org/climate-variability-and-educational-attainment-evidence-from-rural-ethiopia.

4. R. Jisung Park, A. Patrick Behrer, and Joshua Goodman, "Learning Is Inhibited by Heat Exposure, Both Internationally and within the United States," *Nature Human Behavior* 5 (2020), https://www.nature.com/articles/s41562-020-00959-9.

5. Oxfam, "The Hunger Virus: How COVID-19 Is Fuelling Hunger in a Hungry World," Oxfam Media Briefing, July 9, 2020, https://oxfamilibrary.openrepository.com/bitstream/handle/10546/621023/mb-the-hunger-virus-090720-en.pdf.

6. Nicoletta Brazzola and Simon E. M. Helander, *Early Warning Systems* (New York: United Nations Development Programme, 2018), https://www.undp.org/content/dam/rbec/docs/UNDP Brochure Early Warning Systems.pdf.

7. Amanda Taub, "A New Covid-19 Crisis: Domestic Abuse Rises Worldwide," *New York Times*, April 6, 2020, https://www.nytimes.com/2020/04/06/world/coronavirus-domestic-violence.html?referringSource=articleShare.

8. UN Women, *Whose Time to Care? Unpaid Care and Domestic Work During COVID-19* (New York: United Nations, 2020), https://data.unwomen.org/sites/default/files/inline-files/Whose-time-to-care-brief_0.pdf.

9. YOUTH Team, *Youth & COVID-19: Impacts on Jobs, Education, Rights, and Mental Well-being* (Geneva: International Labour Organization, 2020), https://www.ilo.org/global/topics/youth-employment/publications/WCMS_753026/lang--en/index.htm.

10. "Take Action for the Sustainable Development Goals," United Nations, last accessed April 2020, https://www.un.org/sustainabledevelopment/sustainable-development-goals/.

11. *The Sustainable Development Goals Report 2019* (New York: United Nations, 2020), https://unstats.un.org/sdgs/report/2020/The-Sustainable-Development-Goals-Report-2020.pdf.

12. Farah Stockman, "'They're Death Pits': Virus Claims at Least 7,000 Lives in U.S. Nursing Homes," *New York Times*, April 17, 2020, https://www.nytimes.com/2020/04/17/us/coronavirus-nursing-homes.html?referringSource=articleShare.

13. Nishant Kishore et al., "Mortality in Puerto Rico after Hurricane Maris," *New England Journal of Medicine* 379, no. 2 (2018), https://www.nejm.org/doi/full/10.1056/NEJMsa1803972.

14. Simone Weichselbaum, "Hurricane Sandy, One Year Later: Not All Fun and Games in Coney Island," *New York Daily News*, October 26, 2013, https://www.nydailynews.com/new-york/hurricane-sandy/hurricane-sandy-year-coney-island-article-1.1494779.

15. Eric Lipton and Michael Moss, "Housing Agency's Flaws Revealed by Storm," *New York Times*, December 9, 2012, https://www.nytimes.com/2012/12/10/nyregion/new-york-city-housing-agency-was-overwhelmed-after-storm.html.

16. United Nations Office for Disaster Risk Reduction, *Global Assessment Report on Disaster Risk Reduction* (Geneva: United Nations Office for Disaster Risk Reduction, 2019), https://www.undrr.org/publication/global-assessment-report-disaster-risk-reduction-2019.

17. *Experiences of Vulnerable Populations during Disaster: Testimony before the House Transportation and Infrastructure Subcommittee on Economic Development, Buildings, and Emergency Management*, 116th Cong. (2020) (statement of Marcie Roth, World Institute on Disability Executive Director and Chief Executive Officer), https://transportation.house.gov/imo/media/doc/Roth%20Testimony.pdf.

18. History.com Editors, "Hurricane Mitch," History, A&E Television Networks, November 11, 2019, https://www.history.com/topics/natural-disasters-and-environment/hurricane-mitch.

19. Samuel Freije-Rodriguez et al., *Poverty and Shared Prosperity 2020: Reversals of Fortune* (Washington, DC: World Bank Group, 2020), https://openknowledge.worldbank.org/bitstream/handle/10986/34496/9781464816024.pdf.

20. *The Sustainable Development Goals Report 2019* (New York: United Nations, 2019), https://unstats.un.org/sdgs/report/2019/The-Sustainable-Development-Goals-Report-2019.pdf.

21. *The Sustainable Development Goals Report 2019* (New York: United Nations, 2019), https://unstats.un.org/sdgs/report/2019/The-Sustainable-Development-Goals-Report-2019.pdf.

22. Armando Barrientos and David Hulme, "Social Protection for the Poor and Poorest in Developing Countries: Reflections on a Quiet Revolution," *Oxford Development Studies* 37, no. 4 (2009): 439–456, https://www.tandfonline.com/doi/full/10.1080/13600810903305257.
23. Stephane Hallegatte et al., *Unbreakable: Building the Resilience of the Poor in the Face of Natural Disasters* (Washington, DC: World Bank, 2017), https://openknowledge.worldbank.org/handle/10986/25335.
24. L. S. Howard, "Parametric Insurance Can Help Close Global Protection Gap: Clyde & Co. Report," *Insurance Journal*, January 9, 2018, https://www.insurancejournal.com/news/international/2018/01/09/476651.htm.
25. Kathy Baughman McLeod, "Building a Resilient Planet: How to Adapt to Climate Change from the Bottom Up," *Foreign Affairs*, May/June 2020, https://www.foreignaffairs.com/articles/2020-04-13/building-resilient-planet.
26. Danielle Moore et al., *Building Resilience through Financial Inclusion: A Review of Existing Evidence and Knowledge Gaps* (New Haven, CT: Innovations for Poverty Action, 2019), https://www.poverty-action.org/sites/default/files/publications/Building-Resilience-Through-Financial-Inclusion-January-2019.pdf.
27. Anna Hirtenstein, "AXA Insurance Chief Warns of 'Uninsurable Basements' from New York to Mumbai," *Insurance Journal*, January 26, 2018, https://www.insurancejournal.com/news/international/2018/01/26/478615.htm.
28. OECD, *Climate Finance Provided and Mobilised by Developed Countries in 2013–18* (Paris: OECD Publishing, 2020), https://www.oecd-ilibrary.org/docserver/f0773d55-en.pdf?expires=1610643111&id=id&accname=ocid195235&checksum=B81083CF78A45A8B3F11DE772513C21C.
29. Alexander Villegas, Anthony Faiola, and Lesley Wroughton, "As Spending Climbs and Revenue Falls, the Coronavirus Forces a Global Reckoning," *Washington Post*, January 10, 2021, https://www.washingtonpost.com/world/2021/01/10/coronavirus-pandemic-debt-crisis/.
30. Nicole Winfield, "Pope to UN: Use COVID Crisis to Come Out Better, Not Worse," ABC News, September 25, 2020, https://abcnews.go.com/International/wireStory/pope-covid-crisis-worse-73244342.
31. "What Jesus Preached Might Help Us Post-COVID-19," CathNews New Zealand & Pacific, June 22, 2020, https://cathnews.co.nz/2020/

06/22/jesus-debt-jubilee/#:~:text=A%20debt%20jubilee%20On%20
Easter%20Sunday%20Pope%20Francis,at%20least%20some%20
economists%20in%20agreement%20with%20him.

Chapter 5

1. Cumming Cockburn Limited, *Hurricane Hazel and Extreme Rainfall in Southern Ontario* (Toronto: Institute for Catastrophic Loss Reduction, 2000), https://www.iclr.org/wp-content/uploads/PDFS/hurricane-hazel-and-extreme-rainfall-in-southern-ontario.pdf.
2. Cumming Cockburn Limited, *Hurricane Hazel and Extreme Rainfall in Southern Ontario* (Toronto: Institute for Catastrophic Loss Reduction, 2000), https://www.iclr.org/wp-content/uploads/PDFS/hurricane-hazel-and-extreme-rainfall-in-southern-ontario.pdf.
3. DeNederlandscheBank, "Network for Greening the Financial System," Eurosystem, 2019, https://www.dnb.nl/en/about-dnb/co-operation/network-greening-financial-system/index.jsp.
4. Edward Cameron, *Business Adaptation to Climate Change and Global Supply Chains* (Manchester, VT: Global Commission on Adaptation, 2019), https://cdn.gca.org/assets/2019-08/BP%20Supply%20chains_ecameron_Final%2031%20July.pdf.
5. Felix Suntheim and Jérôme Vandenbussche, "Equity Investors Must Pay More Attention to Climate Change Physical Risk," International Monetary Fund, May 29, 2020, https://blogs.imf.org/2020/05/29/equity-investors-must-pay-more-attention-to-climate-change-physical-risk/.
6. Fiona Bayat-Renoux et al., *Tipping or Turning Point: Scaling Up Climate Finance in the Era of COVID-19* (Incheon: Green Climate Fund, 2020), https://www.greenclimate.fund/document/tipping-or-turning-point-scaling-climate-finance-era-covid-19.
7. Eric Roston, "Americans Are Paying $34 Billion Too Much for Houses in Flood Plains," *Claims Journal*, March 3, 2020, https://www.claimsjournal.com/news/national/2020/03/03/295784.htm.
8. Sarah Sinjab, "Bridging the TCFD Gaps," Verisk Maplecroft, May 7, 2020, https://www.maplecroft.com/insights/analysis/bridging-the-tcfd-gaps/.
9. Hon. James Shaw, "New Zealand First in the World to Require Climate Risk Reporting," speech, September 15, 2020, Official website of the New Zealand Government, https://www.beehive.

govt.nz/release/new-zealand-first-world-require-climate-risk-reporting#:~:text=New%20Zealand%20will%20be%20the%20first%20country%20to,mandatory%20across%20the%20financial%20system%2C%E2%80%9D%20James%20Shaw%20said.

10. Michael Bloomberg, "Joe Biden's Big Climate Opportunity," *South Florida Sun Sentinel*, November 13, 2020, https://www.sun-sentinel.com/opinion/commentary/fl-op-com-climate-biden-michael-bloomberg-20201113-scr3termjzdy7af5twwwlpkd3e-story.html.

11. Herman Kahn and Anthony J. Wiener, "The Use of Scenarios," Hudson Institute, last accessed April 2020, https://www.hudson.org/research/2214-the-use-of-scenarios.

12. Herman Kahn and Anthony J. Wiener, "The Use of Scenarios," Hudson Institute, last accessed April 2020, https://www.hudson.org/research/2214-the-use-of-scenarios.

13. Scott Waldman, "Shell Grappled with Climate Change 20 Years Ago, Documents Show," *Scientific American*, Springer Nature, April 5, 2018, https://www.scientificamerican.com/article/shell-grappled-with-climate-change-20-years-ago-documents-show/.

14. Warren G. Bennis, "The Leader: Leading for the Long Run," in *Preparing for a Pandemic* (Cambridge, MA: Harvard Business Review, 2006), https://hbr.org/2006/05/preparing-for-a-pandemic#post5.

15. J. Peter Scoblic and Philip E. Tetlock, "A Better Crystal Ball," *Foreign Affairs*, November/December 2020, https://www.foreignaffairs.com/articles/united-states/2020-10-13/better-crystal-ball.

16. *Final Report of the NSW Bushfire Inquiry* (Sydney: NSW Government, 2020), https://www.dpc.nsw.gov.au/assets/dpc-nsw-gov-au/publications/NSW-Bushfire-Inquiry-1630/Final-Report-of-the-NSW-Bushfire-Inquiry.pdf.

17. Hannah Allam, "Years before the Pandemic, War Games Predicted a 'Global Tempest,'" NPR, May 19, 2020, https://www.npr.org/2020/05/19/853816473/years-before-the-pandemic-war-games-predicted-a-global-tempest.

18. Exec. Order No. 13677, 79 C.F.R. 58229 (September 23, 2014), https://obamawhitehouse.archives.gov/the-press-office/2014/09/23/executive-order-climate-resilient-international-development.

19. I serve as senior advisor to McKinsey.

20. Matt Craven et al., "Not the Last Pandemic: Investing Now to Reimagine Public-Health Systems," McKinsey & Company, July 13, 2020, https://www.mckinsey.com/industries/public-and-social-sector/our-insights/

not-the-last-pandemic-investing-now-to-reimagine-public-health-systems#. I serve as a Senior Advisor to McKinsey & Company.

21. Stephane Hallegatte, "From Pipes to Prosperity: The Resilient Infrastructure Opportunity," World Bank Blogs, July 29, 2019, https://blogs.worldbank.org/climatechange/pipes-prosperity-resilient-infrastructure-opportunity.

22. Jung Eun Oh et al., *Addressing Climate Change in Transport: Volume 2: Pathway to Resilient Transport* (Washington, DC: World Bank, 2019), https://openknowledge.worldbank.org/handle/10986/32412.

23. Stephane Hallegatte, "From Pipes to Prosperity: The Resilient Infrastructure Opportunity," World Bank Blogs, July 29, 2019, https://blogs.worldbank.org/climatechange/pipes-prosperity-resilient-infrastructure-opportunity.

24. Stephane Hallegatte et al., *Lifelines: The Resilient Infrastructure Opportunity* (Washington, DC: World Bank, 2019), https://openknowledge.worldbank.org/bitstream/handle/10986/31805/9781464814303.pdf.

25. Stephane Hallegatte et al., *Lifelines: The Resilient Infrastructure Opportunity* (Washington, DC: World Bank, 2019), https://openknowledge.worldbank.org/bitstream/handle/10986/31805/9781464814303.pdf.

26. Michael Mullan et al., *Climate-Resilient Infrastructure* (Paris: OECD, 2018), http://www.oecd.org/environment/cc/policy-perspectives-climate-resilient-infrastructure.pdf.

27. Anna Moss and Suzanne Martin, *Flexible Adaptation Pathways* (Edinburgh: ClimateXChange, 2012), https://www.climatexchange.org.uk/media/1595/flexible_adaptation_pathways.pdf.

28. "Aim of the Delta Program," Government of the Netherlands, last accessed April 2020, https://www.government.nl/topics/delta-programme/aim-of-the-delta-programme.

29. Blacki Migliozzi et al., "Record Wildfires on the West Coast Are Capping a Disastrous Decade," *New York Times*, September 24, 2020, https://www.nytimes.com/interactive/2020/09/24/climate/fires-worst-year-california-oregon-washington.html.

30. *Final Report of the NSW Bushfire Inquiry* (Sydney: NSW Government, 2020), https://apo.org.au/sites/default/files/resource-files/2020-07/apo-nid307786.pdf.

31. *Final Report of the NSW Bushfire Inquiry* (Sydney: NSW Government, 2020), https://www.dpc.nsw.gov.au/assets/dpc-nsw-gov-au/publications/NSW-Bushfire-Inquiry-1630/Final-Report-of-the-NSW-Bushfire-Inquiry.pdf.

32. *Final Report of the NSW Bushfire Inquiry* (Sydney: NSW Government, 2020), https://www.dpc.nsw.gov.au/assets/dpc-nsw-gov-au/publications/NSW-Bushfire-Inquiry-1630/Final-Report-of-the-NSW-Bushfire-Inquiry.pdf.

33. Anne C. Mulkern, "Utah, Calif. Lead States with New Homes in Fire Zones," *E&E News*, August 28, 2020, https://www.eenews.net/climatewire/2020/08/28/stories/1063712689?utm_medium=email&utm_source=eenews%3Aclimatewire&utm_campaign=edition%2BiZ%2B%2FftFV%2B2LxUfHtN5bxJQ%3D%3D.

34. Kathleen Ronayne, "California Governor Won't Block Building in High-Fire Areas," *U.S. News & World Report*, April 15, 2019, https://www.usnews.com/news/best-states/california/articles/2019-04-15/california-governor-wont-block-building-in-high-fire-areas.

35. *Final Report of the NSW Bushfire Inquiry* (Sydney: NSW Government, 2020), https://www.dpc.nsw.gov.au/assets/dpc-nsw-gov-au/publications/NSW-Bushfire-Inquiry-1630/Final-Report-of-the-NSW-Bushfire-Inquiry.pdf.

36. "$4.2 Trillion Can Be Saved by Investing in More Resilient Infrastructure, New World Bank Report Finds," World Bank, June 19, 2019, https://www.worldbank.org/en/news/press-release/2019/06/19/42-trillion-can-be-saved-by-investing-in-more-resilient-infrastructure-new-world-bank-report-finds.

37. "Volunteerism in Japan: Kobe Earthquake," Trinity College, last accessed April 2020, https://commons.trincoll.edu/japanvolunteerism/great-hanshin-kobe-earthquake/.

38. "A Day to Remember Disasters in Japan," Science Craft, September 1, 2017, https://scraft.co.jp/english/a-day-to-remember-disasters-in-japan-%E9%98%B2%E7%81%BD%E3%81%AE%E6%97%A5%E3%80%81%E6%B4%A5%E6%B3%A2%E9%98%B2%E7%81%BD%E3%81%AE%E6%97%A5/#:~:text=Since%20there%20was%20an%20outpouring,5%20each%20year%20as%20Tsunami.

39. David Segal, "Why Are There Almost No Memorials to the Flu of 1918?," *New York Times*, May 14, 2020, https://www.nytimes.com/2020/05/14/business/1918-flu-memorials.html.

40. John M. Barry, *The Great Influenza: The Story of the Deadliest Pandemic in History* (New York: Penguin Publishing Group, 2005), 394.

41. Russ Banham, "This Insurance Would Have Helped in Coronavirus Crisis but Nobody Bought It," *Insurance Journal*, April 3, 2020, https://

www.insurancejournal.com/news/national/2020/04/03/563224.
htm.

42. United Nations Office for Disaster Risk Reduction (UNDRR),
"Tsunami Countries Share Lessons Learned," UNDRR, October 27,
2017, https://www.undrr.org/news/tsunami-countries-share-lessons-
learned#:~:text.

Chapter 6

1. "How to Grow Green," Bloomberg Green, June 9, 2020, https://www.
bloomberg.com/features/2020-green-stimulus-clean-energy-future/
?sref=Oz9Q3OZU#toaster.

2. Mateo Salazar, *Greenness of Stimulus Index* (London: Vivid Economics,
2020), https://www.vivideconomics.com/wp-content/uploads/2020/
09/GSI_924.pdf.

3. Rico Kongsager, "Addressing Climate Change Mitigation and
Adaptation Together: A Global Assessment of Agriculture and
Forestry Projects," *Environmental Management* 57 (2016): 271–282, https://
link.springer.com/article/10.1007/s00267-015-0605-y.

4. There is lively scholarly debate over the definition of maladaptation.
For example, see, https://www.weadapt.org/knowledge-base/vuln-
ability/maladaptation-an-introduction. For my purposes here, I am
using the term narrowly to reflect the risk of increased global warming
as a result of adaptation.

5. Sarah Mehryar and Swenja Surminski, "National Laws for Enhancing
Flood Resilience in the Context of Climate Change: Potential and
Shortcomings," *Climate Policy* 21, no. 2 (2020), https://www.tandfonline.
com/doi/full/10.1080/14693062.2020.1808439.

6. Molly Walton, "Desalinated Water Affects the Energy Equation
in the Middle East," International Energy Agency, January 21,
2019, https://www.iea.org/commentaries/desalinated-water-
affects-the-energy-equation-in-the-middle-east.

7. "Ghantoot Desalination Pilot Plant," Masdar, 2018, https://masdar.
ae/en/masdar-clean-energy/projects/ghantoot-desalination-
pilot-plant.

8. V. Alexeeva et al., *Climate Change and Nuclear Power 2020*
(Vienna: International Atomic Energy Agency, 2020), https://www-
pub.iaea.org/MTCD/Publications/PDF/PUB1911_web.pdf.

9. Josh Smith, "North Korea Nuclear Reactor Site Threatened by Recent Flooding, U.S. Think-Tank Says," Reuters, August 12, 2020, https://www.reuters.com/article/us-northkorea-nuclear-floods/north-korea-nuclear-reactor-site-threatened-by-recent-flooding-u-s-think-tank-says-idUSKCN25908S.

10. Hannah Ziady, "Solar Power Could Be 'the New King' as Global Electricity Demand Grows," CNN, October 13, 2020, https://edition.cnn.com/2020/10/13/energy/iea-world-energy-outlook-2020/index.html.

11. Jim Malo, "'Snowball Effect': Bushfire Smoke Reduces Solar Panel Efficiency, Increases Load on Coal-Fired Power," Domain, January 15, 2020, https://www.domain.com.au/news/snowball-effect-bushfire-smoke-reduces-solar-panel-efficiency-increases-load-on-coal-fired-power-920909/.

12. Jun Yin, "Impacts of Solar Intermittency on Future Photovoltaic Reliability," *Nature Communications* 11, no. 4781 (2020), https://www.nature.com/articles/s41467-020-18602-6.

13. Jeff Brady, "It's Not Just Texas. The Entire Energy Grid Needs An Upgrade For Extreme Weather," WXXI News, February 28, 2021, https://www.wxxinews.org/post/its-not-just-texas-entire-energy-grid-needs-upgrade-extreme-weather.

14. Peter Behr, "Solar Power Plunges as Smoke Shrouds Calif.," E&E News, September 11, 2020, https://www.eenews.net/energywire/2020/09/11/stories/1063713459?utm_medium=email&utm_source=eenews%3Aenergywire&utm_campaign=edition%2BiZ%2B%2FftFV%2B2LxUfHtN5bxJQ%3D%3D.

15. Pelayo Menéndez et al., "The Global Flood Protection Benefits of Mangroves," *Scientific Reports* 10, no. 4404 (2020), https://doi.org/10.1038/s41598-020-61136-6.

16. Jonathan Sanderman et al., "A Global Map of Mangrove Forest Soil Carbon at 30 m Spatial Resolution," *Environmental Research Letters* 13, no. 5 (2018), https://iopscience.iop.org/article/10.1088/1748-9326/aabe1c.

17. Holly P. Jones et al., "Global Hotspots for Coastal Ecosystem-Based Adaptation," *PLOS One* 15, no. 5 (2020), https://journals.plos.org/plosone/article?id=10.1371/journal.pone.0233005.

18. Michael Tatarski, "A Vital Mangrove Forest Hidden in Vietnam's Largest City Could Be at Risk," Mongabay, April 21, 2020, https://news.mongabay.com/2020/04/a-vital-mangrove-forest-hidden-in-vietnams-largest-city-could-be-at-risk/.

19. Mark E. M. Walton et al., "Are Mangroves Worth Replanting? The Direct Economic Benefits of a Community-Based Reforestation Project," *Environmental Conservation* 33, no. 4 (2006): 335–343, https://www.cambridge.org/core/journals/environmental-conservation/article/abs/are-mangroves-worth-replanting-the-direct-economic-benefits-of-a-community-based-reforestation-project/7D592C07301 0CF5AE64F0407743D36BF.

20. N. Saintilan et al., "Thresholds of Mangrove Survival under Rapid Sea Level Rise," *Science* 368, no. 6495 (2020): 1118–1121, https://science.sciencemag.org/content/368/6495/1118.

21. Ian R. Noble et al., "Adaptation Needs and Options," in *Climate Change 2014: Impacts, Adaptation, and Vulnerability. Part A: Global and Sectoral Aspects. Contribution of Working Group II to the Fifth Assessment Report of the Intergovernmental Panel on Climate Change* (Cambridge: Cambridge University Press, 2014), 844, https://www.ipcc.ch/site/assets/uploads/2018/02/WGIIAR5-Chap14_FINAL.pdf.

22. Rajendra K. Pachauri et al., "Topic 4: Adaptation and Mitigation," in *Climate Change 2014: Synthesis Report, Contribution of Working Groups I, II and III to the Fifth Assessment Report of the Intergovernmental Panel on Climate Change* (Geneva: Intergovernmental Panel on Climate Change, 2015), https://ar5-syr.ipcc.ch/topic_adaptation.php.

23. Dan Asin, "Backdraft: The Conflict Potential of Climate Mitigation and Adaptation," Wilson Center, June 10, 2010, https://www.wilsoncenter.org/event/backdraft-the-conflict-potential-climate-mitigation-and-adaptation.

Conclusion

1. Emily Langer, "Robert M. White, Top Weatherman under Five U.S. Presidents, Dies at 92," *Washington Post*, October 15, 2015, https://www.washingtonpost.com/national/energy-environment/robert-m-white-top-weatherman-under-five-us-presidents-dies-at-92/2015/10/15/79fc3b5e-7291-11e5-9cbb-790369643cf9_story.html?no_nav=true&tid=a_classic-iphone.

2. John H. Cushman Jr., "Shell Knew Fossil Fuels Created Climate Change Risks Back in 1980s, Internal Documents Show," *Inside Climate News*, April 5, 2018, https://insideclimatenews.org/news/05042018/shell-knew-scientists-climate-change-risks-fossil-fuels-global-warming-company-documents-netherlands-lawsuits.

3. John H. Cushman Jr., "Shell Knew Fossil Fuels Created Climate Change Risks Back in 1980s, Internal Documents Show," *Inside Climate News,* April 5, 2018, https://insideclimatenews.org/news/05042018/shell-knew-scientists-climate-change-risks-fossil-fuels-global-warming-company-documents-netherlands-lawsuits.

INDEX

For the benefit of digital users, indexed terms that span two pages (e.g., 52–53) may, on occasion, appear on only one of those pages.

Figures are indicated by *f* following the page number.